T0258589

Molecular Toxinology Handbook

Molecular Toxinology Handbook

Edited by **Cora Lancester**

New York

Published by Callisto Reference,
106 Park Avenue, Suite 200,
New York, NY 10016, USA
www.callistoreference.com

Molecular Toxinology Handbook
Edited by Cora Lancester

© 2015 Callisto Reference

International Standard Book Number: 978-1-63239-469-9 (Hardback)

This book contains information obtained from authentic and highly regarded sources. Copyright for all individual chapters remain with the respective authors as indicated. A wide variety of references are listed. Permission and sources are indicated for detailed attributions, please refer to the permis - sions page. Reasonable efforts have been made to publish reliable data and information, but the authors, editors and publisher cannot assume any responsibility for the validity of all materials or the consequences of their use.

The publisher's policy is to use permanent paper from mills that operate a sustainable forestry policy. Furthermore, the publisher ensures that the text paper and cover boards used have met acceptable environmental accreditation standards.

Trademark Notice: Registered trademark of products or corporate names are used only for explanation and identification without intent to infringe.

Printed in the United States of America.

Contents

Preface

The main aim of this book is to educate learners and enhance their research focus by presenting diverse topics covering this vast field. This is an advanced book which compiles significant studies by distinguished experts in the area of analysis. This book addresses successive solutions to the challenges arising in the area of application, along with it; the book provides scope for future developments.

Molecular Toxinology has been established as a scientific discipline focused on the interconnected description of various aspects of animal toxins, which appear as an invaluable source for the discovery of therapeutic polypeptides in an investigative biotechnological world. Animal toxins rely on particular chemical interactions with their partner molecule to exercise their biological actions. The comprehension of how molecules interact and identify their target is important for the rational exploration of bioactive polypeptides as therapeutics. Investigation on the mechanism of molecular interaction and identification provides a world of new opportunities for the pharmaceutical industry and clinical medicine. This book presents topics on molecular toxinology concerning integration between analytical procedures and biomedical applications.

It was a great honour to edit this book, though there were challenges, as it involved a lot of communication and networking between me and the editorial team. However, the end result was this all-inclusive book covering diverse themes in the field.

Finally, it is important to acknowledge the efforts of the contributors for their excellent chapters, through which a wide variety of issues have been addressed. I would also like to thank my colleagues for their valuable feedback during the making of this book.

Editor

Molecular Toxinology

Toxins from Venomous Animals: Gene Cloning, Protein Expression and Biotechnological Applications

Matheus F. Fernandes-Pedrosa,
Juliana Félix-Silva and Yamara A. S. Menezes

Additional information is available at the end of the chapter

1. Introduction

Venoms are the secretion of venomous animals, which are synthesized and stored in specific areas of their body i.e., venom glands. The animals use venoms for defense and/or to immobilize their prey. Most of the venoms are complex mixture of biologically active compounds of different chemical nature such as multidomain proteins, peptides, enzymes, nucleotides, lipids, biogenic amines and other unknown substances. Venomous animals as snakes, spiders, scorpions, caterpillars, bees, insects, wasps, centipedes, ants, toads and frogs have largely shown biotechnological or pharmacological applications. During long-term evolution, venom composition underwent continuous improvement and adjustment for efficient functioning in the killing or paralyzing of prey and/or as a defense against aggressors or predators. Different venom components act synergistically, thus providing efficiency of action of the components. Venom composition is highly species-specific and depends on many factors including age, sex, nutrition and different geographic regions. Toxins, occurring in venoms and poisons of venomous animals, are chemically pure toxic molecules with more or less specific actions on biological systems [1-3]. A large number of toxins have been isolated and characterized from snake venoms and snake venoms repertoire typically contain from 30 to over 100 protein toxins. Some of these molecules present enzymatic activities, whereas several others are non-enzymatic proteins and polypeptides. The most frequent enzymes in snake venoms are phospholipases A_2, serine proteinases, metalloproteinases, acetylcholinesterases, L-amino acid oxidases, nucleotidases and hyaluronidases. Higher catalytic efficiency, heat stability and resistance to proteolysis as well as abundance of snake venom enzymes provide them attractive models for biotechnologists, pharmacologists and biochemists [3-4]. Scorpion toxins are classified according to their structure, mode of action, and binding site on different channels or channel subtypes. The venom is constituted by mucopolysaccharides, hyaluronidases, phospholipases, serotonins, histamines, enzyme inhibitors, antimicrobials and proteins namely neurotoxic peptides. Scorpion peptides presents specificity and high affinity and have been used as pharmacological tools to characterize various receptor proteins involved in normal ion channel functionating, as abnormal channel functionating in cases of diseases. The venoms can be characterized by identification of peptide toxins analysis of the structure of the toxins and also have proven to be among the most and selective antagonists available for voltage-gated channels permeable to K^+, Na^+, and Ca^{2+}. The neurotoxic peptides and small proteins lead to dysfunction and provoke pathophysiological actions, such as membrane destabilization, blocking of the central, and peripheral nervous systems or alteration of smooth or skeletal muscle activity [5-8]. Spider venoms are complex mixtures of biologically active compounds of different chemical nature, from salts to peptides and proteins. Specificity of action of some spider toxins is unique along with high toxicity for insects, they can be absolutely harmless for members of other taxons, and this could be essential for investigation of insecticides. Several spider toxins have been identified and characterized biochemically. These include mainly ribonucleotide phosphohydrolase, hyaluronidases, serine proteases, metalloproteases, insecticidal peptides and phospholipases D [9-10]. Venoms from toads and frogs have been extensively isolated and characterized showing molecules endowed with antimicrobial and/or cytotoxic activities [11]. Studies involving the molecular repertoire of the venom of bees and wasps have revealed the partial isolation, characterization and biological activity assays of histamines, dopamines, kinins, phospholipases and hyaluronidases. The venom of caterpillars has been partially characterized and contains mainly ester hydrolases, phospholipases and proteases [12]. The purpose of this chapter is to present the main toxins isolated and characterized from the venom of venomous animals, focusing on their biotechnological and pharmacological applications.

2. Biotechnological and pharmacological applications of snake venom toxins

While the initial interest in snake venom research was to understand how to combat effects of snakebites in humans and to elucidate toxins mechanisms, snake venoms have become a fertile area for the discovery of novel products with biotechnological and/or pharmacological applications [13-14]. Since then, many different products have been developed based on purified toxins from snake venoms, as well recent studies have been showing new potential molecules for a variety of applications [15].

2.1. Toxins acting on cardiovascular system

Increase in blood pressure is often a transient physiological response to stressful stimuli, which allows the body to react to dangers or to promptly increase activity. However, when the blood pressure is maintained at high levels for an extended period, its long term effects are highly undesirable. Persistently high blood pressure could cause or accelerate multiple pathological conditions such as organ (heart and kidney) failure and thrombosis events (heart attack and stroke) [14]. So, it is important to lower the blood pressure of high-rick patients through use of specific anti-hypertensive agents, and in this scenario, snake venom toxins has been shown to be promising sources [14-15]. This is because it has long been noted that some snake venoms drastically lower the blood pressure in human victims and experimental animals [15]. The first successful example of developing a drug from an isolated toxin was the anti-hypertensive agent Capoten® (captopril), an angiotensin-converting enzyme (ACE) inhibitor modeled from a venom peptide isolated from *Bothrops jararaca* venom [16]. These bradykinin-potentiating peptides (BPPs) are venom components which inhibits the breakdown of the endogenous vasodilator bradykinin while also inhibiting the synthesis of the endogenous vasoconstrictor angiotensin II, leading to a reduction in blood pressure [15]. BPPs have also been identified in *Crotalus durissus terrificus* venom [17]. Snake venom represents one of the major sources of exogenous natriuretic peptides (NPs) [18]. The first venom NP was identified from *Dendroaspis angusticeps* snake venom and was named *Dendroaspis* natriuretic peptide (DNP) [19]. Other venom NPs were also reported in various snake species, such as *Micrurus corallinus* [20], *B. jararaca* [4], *Trimeresurus flavoviridis*, *Trimeresurus gramineus*, *Agkistrodon halys blomhoffii* [21], *Pseudocerastes persicus* [22], *Crotalus durissus cascavella* [23], *Bungarus flaviceps* [24], among others. L-type Ca^{2+}-channels blockers identified in snake venoms include calciseptine [25] and FS2 toxins [26] from *Dendroaspis polylepis polylepis*, C10S2C2 from *D. angusticeps* [27], S4C8 from *Dendroaspis jamesoni kaimosae* [28] and stejnihagin, a metalloproteinase from *Trimeresurus stejnegeri* [29].

2.2. Toxins acting on hemostasis

Desintegrins are a family of cysteine-rich low molecular weight proteins that inhibits various integrins and that usually contain the integrin-binding RGD motif, that binds the GPIIa/IIIb receptor in platelets, thus prevents the binding of fibrinogen to the receptor and consequently platelet aggregation [13]. Two drugs, tirofiban (Aggrastat®) and eptifibatide (Integrillin®) were designed based on snake venom disintegrins and are avaliable in the market as antiplatelet agents, approved for preventing and treating thrombotic complications in patients undergoing percutaneous coronay intervention and in patients with acute cornonary sydrome [30-31]. Tirofiban has a non-peptide structure mimicking the RDG motif of the disintegrin echistatin from *Echis carinatus* [30]. Eptifibatide is a cyclic peptide based on the KGD motif of barbourin from *Sisturus miliaris barbouri* snake [31]. Recently, leucurogin, a new recombinant disintegrin was cloned from *Bothrops leucurus*, being a potent agent upon platelet aggregation [32]. Thrombin-like enzymes (TLEs) are proteases reported from many different crotalid, viperid and colubrid snakes that share some functional similarity with thrombin [13]. TLEs are not inactivated by heparin-antithrombin III complex (the physiological inhibitor of thrombin), and, differently to thrombin, they are not able to activate FXIII (the enzyme that covalently cross-links fibrin monomer to form insoluble clots). These are interesting properties, because although being procoagulants *in vitro*, TLEs have the clinical results of being anti-coagulants, by the depletion of plasma level of fibrinogen, and the clots formed are easily soluble and removed from the body. At same time, thrombolysis is enhanced by stimulation of endogenous plasminogen activators binding to the noncrosslinked fibrin [13]. Batroxobin (Defibrase®) was isolated and purified from *Bothrops atrox* venom [33] and ancrod (Viprinex®) from *Agkistrodon rhodostoma* [34]. Haemocoagulase® is a mixture of two proteinases isolated from *B. atrox* venom, acting on blood coagulation by two mechanisms: the first having a thrombin-like activity and the second having a thromboplastin-like activity, activating FX which in turn converts prothrombin into thrombin. It is indicated for the prevention and treatment of hemorrhages of a variety of origins [13]. Other toxins acting on hemostasis with potential biotechnological/pharmacological applications has been purified and characterized from several snake venoms, such as bhalternin from *Bothrops alternatus* [35], bleucMP from *B. leucurus* [36], VLH2 from *Vipera lebetina* [37], trimarin from *Trimeresurus malabaricus* [38], BE-I-PLA$_2$ from *Bothrops erythromelas* [39], among others.

2.3. Toxins with antibiotic activity

Antibiotics are a heterogeneous group of molecules produced by several organisms, including bacteria and fungi, presenting an antimicrobial profile, inducing the death of the agent or inhibiting microbial growth [40]. L-amino acid oxidases (LAAOs) are enantioselective flavoenzymes catalyzing the stereospecific oxidative deamination of a wide range of L-amino acids to form α-keto acids, ammonia and hydrogen peroxide (H$_2$O$_2$). Antimicrobial activities are reported to various LAAOs, such as TJ-LAO from *Trimeresurus jerdonii* [41], Balt-LAAO-I from *Bothrops alternatus* [42], TM-LAO *Trimeresurus mucrosquamatus* [43], BpirLAAO-I from *Bothrops pirajai* [44], casca LAO from *Crotalus durissus cascavella* [45], a LAAO from *Naja naja oxiana* [46], BmarLAAO from *Bothrops marajoensis* [47], among others. Recently, studies revealed that *B. jararaca* venom induced programmed cell death in epimastigotes of *Trypanossoma cruzi*, being this anti-*T. cruzi* activity associated with fractions of venoms with LAAO activity [48]. Secreted phospholipases A$_2$ (sPLA$_2$s) constitute a diverse group of enzymes that are widespread in nature, being particularly abundant in snake venoms. In addition to their catalytic activity, hydrolyzing the sn-2 ester bond of glycerophospholipids, sPLA$_2$s display a range of biological actions, which may be either dependent or independent of catalytic action [49]. Eight sPLA$_2$ myotoxins purified from crotalid snake venoms, including both Lys49 and Asp49-type isoforms, were all found to express bactericidal activity [50]. EcTx-I from *Echis carinatus* [51], PnPLA$_2$ from *Porthidium nasutum* [52] and BFPA [53] from *Bungarus fasciatus* also presented antimicrobial activity. Vgf-1, a small peptide from *Naja atra* venom had *in vitro* activity against clinically isolated multidrug-resistant strains of *Mycobacterium tuberculosis* [54]. Neuwiedase, a metalloproteinase from *Bothrops neuwiedi* snake venom, showed considerable effects of *Toxoplasma gondii* infection inhibition *in vitro* [55]. Recently, a study revealed that whole venom, crotoxin and sPLA$_2$s (PLA$_2$-CB and PLA$_2$-IC) isolated from *Crotalus durissus terrificus* venom showed antiviral activity against dengue and yellow fever viruses, which are two of the most important arboviruses in public health [56].

2.4. Toxins acting on inflammatory and nociceptive responses

Various snake venoms are rich in secretory phospholipases A$_2$ (sPLA$_2$), which are potent pro-inflammatory enzymes producing different families of inflammatory lipid mediators such as arachidonic acid derived eicosanoids, various lysophospholipids and platelet activating factors through cyclooxygenase and lipoxygenase pathways [57]. In a recent study, was described the first complete nucleotide sequence of a βPLI from venom glands of *Lachesis muta* by a transcriptomic analysis [58]. Recently, was purified from the venom of *Crotalus durissus terrificus* a hyaluronidase (named Hyal) that was able to provide a highly antiedematogenic acitivity [59]. Crotapotin, a subunit of crotoxin, from *C. d. terrificus*, has been reported to possess immunossupressive activity, associated to an increase in the production of prostaglandin E_2 by macrophages, consequently reducing the proliferative response of lymphocytes [60]. Various elapid and viperid venoms have been reported to induce antinociception through their neurotoxins and myotoxins [61]. *C. d. terrificus* venom induces neurological symptoms in their victims, but, contrary to most venoms from other species, it does not induce pain or severe tissue destruction at the site of inoculation, being usual the sensation of paresthesia in the affected area [62]. Based on this, several studies have been carried out with this venom, being reported in the literature several molecules with antinociceptive activity from *C. d. terrificus* venom, such as crotamine [63] and crotoxin [64]. Has been demonstrated that the anti-nociceptive effect of crotamine involve both central and peripheral mechanisms, being 30-fold higher than the produced by morphine [63]. Studies suggest that crotoxin has antinociceptive effect mediated by an action on the central nervous system, without involvement of muscarinic and opioid receptors [64]. Other antinociceptive peptides isolated from snake venoms are cobrotoxin, a neurotoxin isolated from *Naja atra* [65] and hannalgesin, a neurotoxin isolated from *Ophiophagus hannah* [66].

2.5. Toxins acting on immunological system

Venom-derived peptides are being evaluated as immunosuppressants for the treatment of autoimmune diseases and the prevention of graft rejection [67]. Studies have shown that anti-crotalic serum possesses an antibody content usually inferior to the antibody content of other anti-venom serum suggesting that the crotalic venom is a poor immunogen or that it has components with immunosuppressor activity [68]. Indeed, the immunosuppressive effect of venom and crotoxin (a toxin isolated from *Crotalus durissus terrificus*) was reported [68]. Crotapotin, an acidic and non-toxic subunit of crotoxin, administered by intraperitoneal route, significantly reduces the severity of experimental autoimmune neuritis, an experimental model for Guillain-Barré syndrome, which indicate a novel path for neuronal protection in this autoimmune disease and other inflammatory demyelinating neuropathies [69]. Inappropriate activation of complement system occurs in a large number of inflammatory, ischaemic and other diseases. Cobra venom factor (CVF) is an unusual venom component which exists in the venoms of different snake species, such as *Naja* sp., *Ophiophagus* sp. and *Hemachatus* sp. that activate complement system [70]. Due its similarity with C3 complement system component, after binding to mammalian fB in plasma and cleavage of fB by fD, produces a C3 convertase, that is more stable than the other C3 convertases, and resistant to the fluid phase regulators. The CVF-Bb convertase consumes all plasma C3 obliterating the functionality of complement system [70]. Recently, a CVF named OVF was purified from the crude venom of *Ophiophagus hannah* and cloned by cDNA transcriptomic analysis of the snake venom glands [71].

2.6. Toxins with anticancer and cytotoxic activities

Anticancer therapy is an important area for the application of proteins and peptides from venomous animals. Integrins play multiple important roles in cancer pathology including tumor cell proliferation, angiogenesis, invasion and metastasis [72]. Inhibition of angiogenesis is one of the heavily explored treatment options for cancer, and in this scenario snake venom disintegrins represent a library of molecules with different structure, potency and specificity [1]. RGD-containing disintegrins was identified in several snake venoms, inhibiting tumor angiogenesis and metastasis, such as accutin (from *Agkistrodon acutus*) [73], salmosin (from *Agkistrodon halys brevicaudus*) [74], contortrostatin (from *Agkistrodon contortrix*) [75], jerdonin (from *Trimeresurus jerdonii*) [76], crotatroxin (from *Crotalus atrox*) [77], rhodostomin (from *Calloselasma rhodostoma*) [78] and a novel desintegrin from *Naja naja* [79]. The cytostatic effect of L-amino acid oxidases (LAAOs) have been demonstrated using various models of human and animal tumors. Studies show that LAAOs induces apoptosis in vascular endothelial cells and inhibits angiogenesis [80]. Examples of LAAOs isolated from snake venoms with anticancer potential are a LAAO isolated from *Ophiophagus hannah* [81], ACTX-6 from *A. acutus* [82], OHAP-1 from *Trimeresurus flavoviridis* [83] and Bl-LAAO from *Bothrops leucurus* [84]. Secretory phospholipases A$_2$ (sPLA$_2$) also figures the snake toxins with anticancer potential [1]. sPLA$_2$ with cytotoxic activity to tumor cells was described in *Bothrops neuwiedii* [85], *Bothrops brazili* [86], *Naja naja naja* [87], among others. Crotoxin, the main polypeptide isolated from *C. d. terrificus* has shown potent antitumor activity as well the whole venom, highlighting thereby the potential of venom as a source of pharmaceutical templates for cancer therapy [88]. BJcuL, a lectin purified from *Bothrops jararacussu* venom [89] and a metalloproteinase [90] and a lectin from *B. leucurus* [91] are other examples of toxins from snake venoms with anticancer potential.

3. Biotechnological and pharmacological applications of scorpion venom toxins

Scorpions are venomous arthropods, members of Arachnida class and order Scorpiones. These animals are found in all continents except Antarctica, and are known to cause problems in tropical and subtropical regions. Actually these animals are represented by 16 families and approximately 1500 different species and subspecies which conserved their morphology almost unaltered [92-93]. The scorpion species that present medically importance belonging to the family Buthidae are represented by the genera *Androctonus, Buthus, Mesobuthus, Buthotus, Parabuthus,* and *Leirus* located in North Africa, Asia, the Middle East, and India. *Centruroides* spp. are located in Southwest of United States, Mexico, and Central America, while *Tityus* spp. are found in Central and South America and Caribbean. In these different regions of the world the scorpionism is considered a public health problem, with frequent statements that scorpion stings are dangerous [8]. It is generally known that scorpion venom is a complex mixture composed of a wide array of substances. It contains mucopolysaccharides, hyaluronidase, phopholipase, low relative molecular mass molecules like serotonin and histamine, protease inhibitors, histamine releasers and polypeptidyl compounds. Scorpion venoms are a particularly rich source of small, mainly neurotoxic proteins or peptides interacting specifically with various ionic channels in excitable membranes [94].

3.1. Toxins acting on cardiovascular system

The first peptide from scorpion endowed effects of bradykinin and on arterial blood pressure was isolated from the Brazilian scorpion *Tityus serrulatus* [95]. These peptides named *Tityus serrulatus* Hypotensins have molecular masses ranging approximately from 1190 to 2700 Da [96]. Other scorpion bradykinin-potentiating peptides (BPPs) were reported to be found in the venom of the scorpions *Buthus martensii* Karsch [97] and *Leiurus quinquestriatus* [98]. These molecules can display potential as new drugs and could be of interest for biotechnological purposes.

3.2. Toxins with antibiotic activity

In order to defend themselves against the hostile environment, scorpions have developed potent defensive mechanisms that are part of innate and adaptive immunity [99]. Cysteine-free antimicrobial peptides have been identified and characterized from the venom of six scorpion species [100]. Antimicrobial peptides isolated from scorpion venom are important in the discovery of novel antibiotic molecules [101]. The first antimicrobial peptide isolated from scorpions were of the defensin type from *Leiurus quinquestriatus hebraeus* [102]. Later cytolitic and/or antibacterial peptides were isolated from scorpions belonging to the Buthidae, Scorpionidae, Ischnuridae, and Iuridae superfamilies hemo-lymph and venom [103-108]. The discovery of these peptides in venoms from Eurasian scorpions, Africa and the Americas, confirmed their widespread occurrence and significant biological function. Scorpine, a peptide from *Pandinus imperator* with 75 amino acids, three disulfide bridges, and molecular mass of 8350 Da has anti-bacterial and anti-malaria effects [104]. A cationic amphipatic peptide consisting of 45 amino acids has been purified from the venom of the southern African scorpion, *Parabuthus schlechteri*. At higher concentrations it forms non-selective pores into membranes causing depolarization of the cells [109]. Opistoporin1 and 2 (OP 1 and 2) was isolated from the venom of *Opistophthalmus carinatus*. These are amphipathic, cationic peptides which differ only in one amino acid residue. OP1 and PP were active against Gram-negative bacteria and both had hemolytic activity and antifungal activity. These effects are related to membrane permeabilization [106]. A new antimicrobial peptide, hadrurin, was isolated from *Hadrurus aztecus*. It is a basic peptide composed of 41 amino-acid residues with a molecular mass of 4436 Da, and contains no cysteines. It is a unique peptide among all known antimicrobial peptides described, only partially similar to the N-terminal segment of gaegurin 4 and brevinin 2e, isolated from frog skin. It would certainly be a model molecule for studying new antibiotic activities and peptide-lipid interactions [110]. Pandinin 1 and 2 are antimicrobial peptides have been identified and characterized from venom of the African scorpion *Pandinus imperator* [101]. Recently six novel peptides, named bactridines, were isolated from *Tityus discrepans* scorpion venom by mass spectrometry. The antimicrobial effects on membrane Na$^+$ permeability induced by bactridines were observed on *Yersinia enterocolitica* [111]. The profile of gene in the venom glands of *Tityus stigmurus* scorpions was studied by transcriptome. Data revealed that 41 % of ESTs belong to recognized toxin-coding sequences, with transcripts encoding antimicrobial toxins (AMP-like) being the most abundant, followed by alfa KTx-like, beta KTx-like, beta NaTx-like and alfa NaTx-like. Parallel, 34% of the transcripts encode "other possible venom molecules", which correspond to anionic peptides, hypothetical secreted peptides, metalloproteinases, cystein-rich peptides and lectins [7].

3.3. Toxins acting on acting on inflammatory and nociceptive response

The use of toxins as novel molecular probes to study the structure-function relationship of ion-channels and receptors as well as potential therapeutics in the treatment of wide variety of diseases is well documented. The high specificity and selectivity of these toxins have attracted a great deal of interest as candidates for drug development [8]. At least five peptides have been identified from *Buthus martensii* (Chinese scorpion) venom that have anti-inflammatory and antinociceptive properties [61]. One peptide, J123, blocks potassium channels that activate memory T-cells [112]. The venom also contains a 61-amino acid peptide that has demonstrated antiseizure properties in an animal model [113] as well as other constituents that act as analgesics in mice, rats, and rabbits [114]. The polypeptide BmK IT2 from scorpion *Buthus martensi* Karsh stops rats from reacting to experimentally-induced pain [115]. A protein from the Indian black scorpion, *Heterometrus bengalensis*, bengalin caused human leukemic cells to undergo apoptosis *in vitro* [116]. The peptide chlorotoxin, found in the venom of the scorpion *Leiurus quinquestriatus*, retarded the activity of human glioma cells *in vitro* [117]. An investigation about the role of kinins, prostaglandins and nitric oxide in mechanical hypernociception, spontaneous nociception and paw oedema after intraplantar have been done with *Tityus serrulatus* venom in male wistar rats, proving the potential of use of the venom to alleviate pain and oedema formation [118].

3.4. Toxins acting on acting on immunological system

OSK1 (alpha-KTx3.7) is a 38-residue toxin cross-linked by three disulphide bridges initially purified from the venom of the central Asian scorpion *Orthochirus scrobiculosus* [119]. OSK1 and several structural analogues were produced by solid-phase chemical synthesis, and were tested for lethality in mice and for their efficacy in blocking a series of 14 voltage-gated and Ca^{2+} activated K^+ channels *in vitro*. The literature report that OSK1 could serve as leads for the design and production of new immunosuppressive drugs [119]. Margatoxin, a peptidyl inhibitor of K^+ channels has been purified to homogeneity from venom of the new world scorpion *Centruroides margaritatus* showed that could be used as immunosuppressive agent [120]. Kaliotoxin, a peptidyl inhibitor of the high conductance Ca^{2+}-activated K^+ channels (KCa) has been purified to homogeneity from the venom of the scorpion *Androctonus mauretanicus mauretanicus*. This peptide appears to be a useful tool for elucidating the molecular pharmacology of the high conductance Ca^{2+}-activated K^+ channel [121]. Agitoxin 1, 2, and 3, from the venom of the scorpion *Leiurus quinquestriatus* var. hebraeus have been identified on the basis of their ability to block the shaker K^+ channel [122]. Hongotoxin, a peptide inhibitor of shaker-type (K(v)1) K⁺ channels have been purified to homogeneity from venom of the scorpion *Centruroides limbatus* [123]. Noxiustoxin, component II-11 from the venom of scorpion *Centruroides noxius* Hoffmann, was obtained in pure form after fractionation by Sephadex G-50 chromatography followed by ion exchange separation on carboxymethylcellulose columns. This peptide is the first short toxin directed against mammals and the first K⁺ channel blocking polypeptide-toxin found in scorpion venoms [124]. Pi1 is a peptide purified and characterized from the venom of the scorpion *Pandinus imperato*, showing ability to block the shaker K⁺ channel [125]. All of these peptides obtained from scorpions venoms are potential toxins acting on immunological system as immunosuppressant for autoimmune diseases.

3.5. Toxins with anticancer and cytotoxic activities

One of the most notable active principles found in scorpion venom is chlorotoxin (Cltx), a peptide isolated from the species *Leiurus quinquestriatus*. Cltx has 36 amino acids with four disulfide bonds, and inhibits chloride influx in the membrane of glioma cells [126]. This peptide binds only to glioma cells, displaying little or no activity at all in normal cells. The toxin appears to bind matrix metalloproteinase II [117]. A synthetic version of this peptide (TM601) is being produced by the pharmaceutical industry coupled to iodine 131 (131I-TM601), to carry radiation to tumor cells [127]. A recent study shows that TM601 inhibited angiogenesis stimulated by pro-angiogenic factors in cancer cells, and when TM601 was co-administered with bevacizumab, the combination was significantly more potent than a ten-fold increase in bevacizumab dose [128]. A chlorotoxin-like peptide has also been isolated, cloned and sequenced from the venom of another scorpion species, *Buthus martensii* Karsch [129]. In reference [130] was expressed the recombinant chlorotoxin like peptide from *Leiurus quinquestriatus* and named rBmK CTa. Two novel peptides named neopladine 1 and neopladine 2 were purified from *Tityus discrepans* scorpion venom and found to be active on human breast carcinoma SKBR3 cells. Inmunohistochemistry assays revealed that neopladines bind to SKBR3 cell surface inducing FasL and BcL-2 expression [131]. Results indicate the venom from this scorpion represents a great candidate for the development of new clinical treatments against tumors.

3.6. Toxins with insecticides applications

Evidence for the potential application of scorpions toxins as insecticides has emerged in recent years. The precise action mechanism of several of these molecules remains unknown; many have their effects via interactions with specific ion channels and receptors of neuromuscular systems of insects and mammals. These highly potent and specific interactions make venom constituents attractive candidates for the development of novel therapeutics, pesticides and as molecular probes of target molecules [132].

Toxin Lqhα IT from the scorpion *Leiurus quinquestriatus hebraeus* venom is the best representative of anti-insect alpha toxins [133-134]. A similar effect was observed after applying the insect-selective toxin Bot IT1 from *Buthus occitanus tunetanus* venom [135]. Selective inhibition of the inactivation process of the insect para/tipNav expressed in *Xenopus oocyteswas*

was observed in the presence of Bjα IT [136] and OD1 [137], which are toxins from *Buthotus judaicus* and *Odonthobuthus doriae* scorpion venom, respectively. A second group of scorpion toxins slowing insect sodium channel inactivation was called alpha-like toxins. The first precisely described toxins from this group were the Lqh III/Lqh3 (from *L. q. hebraeus*), Bom III/ Bom 3 and Bom IV/ Bom 4 (from *B. o. mardochei*). They were all tested on cockroach axonal preparation [138-139]. BmKM1 toxin from *B. martensi* Karsch was the first alpha-like toxin available in recombinant form that was tested also on cockroach axonal preparation [140]. Toxins Lqh6 and Lqh7 from *L. q. hebraeus* scorpion venom show high structural similarity with Lqh3 toxin. Their toxicity to cockroach is in the range found for other alpha-like toxins [141]. Alpha-like toxins from scorpion venoms show lower efficiency when applied to insects, as compared to α anti-insect toxins. Therefore they seem to be less interesting from the point of view of future insecticide development [132]. Scorpion contractive and depressant toxins are highly selective for insect sodium channels. Several of these toxins were tested on cockroach axonal preparations; toxin AaI I IT1 from the *A. australis* scorpion venom was the first one [142-143]. All other contractive toxins tested on cockroach axon produced very similar effects, as for example Lqq IT1 from *L. q. quinquestriatus* [133]; Bj IT1 from *B. judaicus* [143], Bm 32-1 and Bm 33-1 from *B. martensi* [144].

4. Biotechnological and pharmacological applications of spider venom toxins

Spider venoms contain a complex mixture of proteins, polypeptides, neurotoxins, nucleic acids, free amino acids, inorganic salts and monoamines that cause diverse effects in vertebrates and invertebrates [145]. Regarding the pharmacology and biochemistry of spider venoms, they present a variety of ion channel toxins, novel non-neurotoxins, enzymes and low molecular weight compounds [146].

4.1. Toxins acting on cardiovascular system

Venom from the South American tarantula *Grammostola spatulata* presents GsMtx-4, a small peptide belonging to the "cysteine-knot" family that blocks cardiac stretch-activated ion channels and suppresses atrial fibrillation in rabbits [147]. Studies are being conducted to develop therapeutics for atrial fibrillation based on GsMtx-4.

4.2. Toxins acting on hemostasis

ARACHnase (Hemostasis Diagnostics International Co., Denver, CO) is a normal plasma that contains a venom extract from the brown recluse spider, *Loxosceles reclusa*, which mimics the presence of a lupus anticoagulant (LA). ARACHnase is a biotechnological product usefulness like a positive control for lupus anticoagulant testing [148]. Native dermonecrotic toxins (phospholipase-D) from *Loxosceles* sp. are agents that stimulate platelet aggregation [149].

4.3. Toxins with antibiotic activity

Two peptide toxins with antimicrobial activity, lycotoxins I and II, were identified from venom of the wolf spider *Lycosa carolinensis* (Araneae: Lycosidae). The lycotoxins may play a dual role in spider-prey interaction, functioning both in the prey capture strategy as well as to protect the spider from potentially infectious organisms arising from prey ingestion. Spider venoms may represent a potentially new source of novel antimicrobial agents with important medical implications [150].

4.4. Toxins acting on inflammatory and nociceptive response

Psalmotoxin 1, a peptide extracted from the South American tarantula *Psalmopoeus cambridgei*, has very potent analgesic properties against thermal, mechanical, chemical, inflammatory and neuropathic pain in rodents. It exerts its action by blocking acid-sensing ion channel 1a, and this blockade results in an activation of the endogenous enkephalin pathway [151]. Phospholipases from both *Loxosceles laeta* and *Loxosceles reclusa* cleaved LPC (lysophosphatidylcholine) to LPA (lysophosphatidic acid) and choline. LPA receptors are potential targets for *Loxosceles* sp. envenomation treatment [152]. The possibilities for biotechnological applications in this area are enormous. Recombinant dermonecrotic toxins could be used as reagents to establish a new model to study the inflammatory response, as positive inducers of the inflammatory response and edema [9, 153-154]. The phospholipase-D from *Loxosceles* venom could be used in phospholipid studies, specially studies on cell membrane constituents with emphasis upon sphingophospholipids, lysophospholipids, lysophosphatidic acid and ceramide-1-phosphate, as models for elucidating lipid product receptors, signaling pathways and biological activities; this new wide field of *Loxosceles* research could also reveal new targets for the treatment of envenomation [10].

4.5. Toxins acting on immunological system

The antiserum most commonly used for treatment of loxoscelism in Brazil is anti-arachnidic serum. This serum is produced by the Instituto Butantan (São Paulo, Brazil) by hyperimmunization of horses with venoms of the spiders *Loxosceles gaucho* and *Phoneutria nigriventer* and the scorpion *Tityus serrulatus*. Several studies have indicated that sphingomyelinase D (SMase D) in venom of *Loxosceles* sp. spiders is the main component responsible for local and systemic effects observed in loxoscelism [153, 155]. Neutralization tests showed that anti-SMase D serum has a higher activity against toxic effects of *L. intermedia* and *L. laeta* venoms and similar or slightly weaker activity against toxic biological effects of *L. gaucho* than that of Arachnidic serum. These results demonstrate that recombinant SMase D can replace venom for anti-venom production and therapy [155].

4.6. Toxins with anticancer and cytotoxic activities

Psalmotoxin 1 was evaluated on inhibited Na^+ currents in high-grade human astrocytoma cells (glioblastoma multiforme, or GBM). These observations suggest this toxin may prove useful in determining whether GBM cells express a specific ASIC-containing ion channel type that can serve as a target for both diagnostic and therapeutic treatments of aggressive malignant gliomas [156]. The antitumor activity of a potent antimicrobial peptide isolated from hemocytes of the spider *Acanthoscurria gomesiana*, named gomesin, was tested *in vitro* and *in vivo*. Gomesin showed cytotoxic and antitumor activities in cell lines, such as melanoma, breast cancer and colon carcinoma [157].

4.7. Toxins with insecticides applications

Several spider toxins have been studied as potential insecticidal bioactive with great biotechnological possible applications [10]. A component of the venom of the Australian funnel web spider *Hadronyche versuta* that is a calcium channel antagonist retains its biological activity when expressed in a heterologous system. Transgenic expression of this toxin in tobacco effectively protected the plants from *Helicoverpa armigera* and *Spodoptera littoralis* larvae, with 100% mortality within 48h [158]. LiTxx1, LiTxx2 and LiTxx3 from *Loxosceles intermedia* venom were identified containing peptides that were active against *Spodoptera frugiperda*. These venom-derived products open a source of insecticide toxins that could be used as substitutes for chemical defensives and lead to a decrease in environmental problems [159]. An insecticidal peptide referred to as Tx4(6-1) was purified from the venom of the spider *Phoneutria nigriventer* by a combination of gel filtration, reverse-phase fast liquid chromatography on Pep-RPC, reverse-phase high performance liquid chromatography (HPLC) on Vydac C18 and ion-exchange HPLC. The protein contains 48 amino acids including 10 Cys and 6 Lys. The results showed that Tx4(6-1) has no toxicity for mice, and suggest that it is a specific anti-insect toxin [160]. SMase D and homologs in the SicTox gene family are the most abundantly expressed toxic protein in venoms of *Loxosceles* and *Sicarius* spiders (Sicariidae). A recombinant SMase D from *Loxosceles arizonica* was obtained and compared its enzymatic and insecticidal activity to that of crude venom. SMase D and crude venom have comparable and high potency in immobilization assays on crickets. These data indicate that SMase D is a potent insecticidal toxin, the role for which it presumably evolved [161]. δ-PaluIT1 and δ-paluIT2 are toxins purified from the venom of the spider *Paracoelotes luctuosus*. Similar in sequence to μ-agatoxins from *Agelenopsis aperta*, their pharmacological target is the voltage-gated insect sodium channel, of which they alter the inactivation properties in a way similar to α-scorpion toxins. Electrophysiological experiments on the cloned insect voltage-gated sodium channel heterologously co-expressed with the tipE subunit in *Xenopus laevis* oocytes, that δ-paluIT1 and δ-paluIT2 procure an increase of Na^+ current [162]. Recently, several toxins have been isolated from spiders with potential biotechnological application as insecticide.

5. Biotechnological and pharmacological applications of toad and frog toxins

Amphibians (toads, frogs, salamanders etc.) during their evolution have developed skin glands covering most parts of their body surface. From these glands small amounts of a mucous slime are secreted permanently, containing substances with different pharmacologic activities such as cardiotoxins, neurotoxins, hypotensive as well as hypertensive agents, hemolysins, and many others. Chemically they belong to a wide variety of substance classes such as steroids, alkaloids, indolalkylamines, catecholamines and low molecular peptides [11, 163]. Several studies have been showing new potential molecules for a variety of pharmacological applications from toads and frogs venoms.

5.1. Toxins acting on cardiovascular system

Neurotensin-like peptides has been identified from frog skin, such as margaratensin, isolated from *Rana margaratae* [164], a potential antihypertensive drug. Similar to the cardiac glycosides, bufadienolides from *Bufo bufo gargarizans* toad skin are able of inhibiting Na^+/K^+-ATPase, having an important role on treatment of congestive heart failure and arterial hypertension [165]. Examples of these bufadienolides are arenobufagin [166], cinobufagin, bufalin, resibufogenin, among others [165]. In the skin of *Rana temporaria* and *Rana igromaculata* frogs, bradykinin, a hypotensive and smooth muscle exciting substance, has been found [11]. Atelopidtoxin, a water-soluble toxin from skin of *Atelopus zeteki* frog, when injected into mammals, produces hypotension and ventricular fibrillation [167]. Semi-purified skin extracts from *Pseudophryne coriacea* frog displayed effects on systemic blood pressure, reducing it by a probably cholinergic mechanism [168].

5.2. Toxins acting on hemostasis

Annexins are a well-known multigene family of Ca^{2+}-regulated membrane-binding and phospholipid-binding proteins. A novel annexin A2 (Bm-ANXA2) was isolated and purified from *Bombina maxima* skin homogenate, being the first annexin A2 protein reported to possess platelet aggregation-inhibiting activity [169].

5.3. Toxins with antibiotic activity

Toxins with antibiotic activity are the most well studied toxins in toads and frogs. Two antimicrobial bufadienolides, telocinobufagin and marinobufagin, were isolated from skin secretions of the Brazilian toad *Bufo rubescens* [170]. Antimicrobial peptides, named syphaxins (SPXs), were isolated from skin secretions of *Leptodactylus syphax* frog [171]. The alkaloids apinaceamine, 6-methyl-spinaceamine isolated from the skin gland secretions of *Leptodactylus pentadactylus* showed in screening tests bactericidal activity [172]. The cinobufacini and its active components bufalin and cinobufagin, from *Bufo bufo gargarizans* Cantor skin, presented anti-hepatitis B virus (HBV) activity [173]. Telocinobufagin from *Rhinella jimi* toad were demonstrated to be active against *Leishmania chagasi* promastigotes and *Trypanosoma cruzi* trypomastigotes, while hellebrigenin, from same source, was active against only *T. cruzi* trypomastigotes [174].

5.4. Toxins acting on inflammatory and nociceptive responses

Epibatidine, an azabicycloheptane alkaloid isolated from the skin of frog *Epipedobates tricolor*, was found to be a potent antinociceptive compound. Although its toxicity, this toxin could be a lead compound in the development of therapeutic agents for pain relief as well for treatment of disorders whose pathogenesis involves nicotinic receptors [175]. A variety of toxins acting on opioid receptors have been isolated from amphibians. Dermorphin (Tyr-D-Ala-Phe-Gly-Tyr-Pro-Ser-NH$_2$) and related heptapeptide [Hyp^6]-dermorphin isolated from the frog skin of *Phyllomedusa* sp., show higher affinity for μ-opioid receptors. Several peptides belonging to the dermorphin family have been isolated from frog skin [61]. Deltorphins (also referred as dermenkephalin) and related peptides isolated from the frog skin have been found to exhibit high selectivity for δ-opiate receptors [176].

5.5. Toxins with anticancer and cytotoxic activities

Venenum Bufonis is a traditional Chinese medicine obtained from the dried white secretion of auricular and skin glands of Chinese toads (*Bufo melanostictus* Schneider or *Bufo bufo gargarzinas* Cantor). Cinobufagin (CBG), isolated from *Venenum Bufonis*, had potential immune system regulatory effects and is suggested that this compound could be developed as a novel immunotherapeutic agent to treat immune-mediated diseases such as cancer [177]. Bufadienolides from toxic glands of toads are used as anticancer agents, mainly on leukemia cells. Bufalin and cinobufagin from *Bufo bufo gargarizans* Cantor were tested and studies shown that these toxins suppress cell proliferation and cause apoptosis in prostate cancer cells via a sequence of apoptotic modulators [178]. Bufotalin, one of the bufadienolides isolated from Formosan Ch'an Su, which is made of the skin and parotid glands of toads, induce apoptosis in human hepatocellular carcinoma, probably involving caspases and apopotosis-inducing factor [179]. Cutaneous venom of *Bombina variegata pachypus* toad presented a cytolitic effect on the growth of the human HL 60 cell line [180]. Brevinin-2R, a non-hemolytic defensin has been isolated from the skin of the frog *Rana ridibunda*, showing pronounced cytotoxicity towards malignant cells [181].

5.6. Toxins with insulin releasing activity

Diabetes mellitus is a disease in which the body is unable to sufficiently produce or properly use insulin. Newer therapeutic modalities for this disease are extremely needed. Peptides with insulin-releasing activity have been isolated from the skin secretions of the frog *Agalychnis litodryas* and may serve as templates for a novel class of insulin secretagogues [182].

6. Biotechnological and pharmacological applications of bee and wasp toxins

Stinging accidents caused by wasps and bees generally produce severe pain, local damage and even death in various vertebrates including man, caused by action of their venoms. Bee venom contains a variety of compounds peptides including melittin, apamin, adolapin, and mast cell degranulating (MCD) peptide, in addition of hyaluronidase and phospholipase A enzymes, that plays a variety of biological activities. The chemical constituents of venoms from wasps species include acetylcholine, serotonin, norepinephrine, hyaluronidase, histidine decarboxylase, phospholipase A_2 and several polycationic peptides and proteins [12].

6.1. Toxins acting on cardiovascular system

Honey bee venom and its main constituents have a marked effect on the cardiovascular system, most notably a fall in arterial blood pressure [183]. From the hemodynamic point of view, the venom, in higher doses, is extremely toxic to the circulatory system and in smaller doses, however, produce a stimulatory effect upon the heart [184]. Melittin, a strongly basic 26 amino-acid polypeptide which constitutes 40–60% of the whole dry honeybee venom, induces contractures and depolarization in skeletal muscle [12]. Melittin is cardiotoxic *in vitro*, causing arrest of the rat heart, but only induces a slight hypertension *in vivo* [183]. Apamin, without direct effect on contraction or relaxation, could attenuate the relaxation evoked by melittin at lower concentrations, and thus contribute to the conversion of melittin's relaxing activity into the contractile activity of the venom. Another peptide found in bee venom that outlines effects on the cardiovascular system is the Cardiopep. Cardiopep is a relatively nonlethal component, compared to phospholipase A, melittin, or whole bee venom itself. It is a potent nontoxic beta-adrenergio-like stimulant that possesses definite anti-arrhythmic properties [185]. Studies on the cardiovascular effects of mastoparan B, isolated from the venom of the hornet *Vespa basalis*, has shown that the peptide caused a dose-dependent inhibition of blood pressure and cardiac function in the rat. Research has shown that the cardiovascular effects of mastoparan B are mainly due to the actions of serotonin, and by a lesser extent to other autacoids, released from mast cells as well from other biocompartments [186].

6.2. Toxins acting on hemostasis

The mechanism by which bee venom affects the hemostatic system remains poorly understood [187]. Among the serine proteases isolated from bees, which acts as a fibrin(ogen)olytic enzyme, activator prothrombin and directly degrades fibrinogen into fibrin degradation products, are the Bi-VSP (*Bombus ignitus*) [188], Bt-VSP (*Bombus terrestris*) [189] and Bs-VSP (*Bombus hypocrita sapporoensis*) [190]. According reference [188], the activation of prothrombin and fibrin(ogen)olytic activity may cooperate to effectively remove fibrinogen, and thus reduce the viscosity of blood. The injection fibrin(ogen)olytic enzyme can be used to facilitate the propagation of components of bee venom throughout the bloodstream of mammals. Bumblebee venom also affects the hemostatic system through by Bi-KTI (*B. ignitus*), a Kunitz-type inhibitor, that strongly inhibited plasmin during fibrinolysis, indicating that Bi-KTI specifically targets plasmin [187]. A toxin protein named magnvesin was purified of *Vespa magnifica*. This protein contains serine protease-like activity inhibits blood coagulation, and was found to act on factors TF, VII, VIII, IX and X [191]. Other anticoagulant protein (protease I) with proteolytic activity was purified from *Vespa orientalis* venom, involving mainly coagulation factors VIII and IX [192]. Magnifin, a phospholipase A_1 (PLA$_1$) purified from wasp venoms of *V. magnifica*, is very similar to other (PLA$_1$), especially to other wasp allergen PLA$_1$. Magnifin can activate platelet aggregation and induce thrombosis *in vivo*. It was the first report of PLA$_1$ from wasp venoms that can induce platelet aggregation [193].

6.3. Toxins with antibiotic activity

Antimicrobial peptides have attracted much attention as a novel class of antibiotics, especially for antibiotic-resistant pathogens. They provide more opportunities for designing novel and effective antimicrobial agents [194]. Melittin has various biological, pharmacological and toxicological actions including antibacterial and antifungal activities [195]. Bombolitin (structural and biological properties similar to those of melittin), isolated from the venom of *B. ignitus* worker bees, possesses antimicrobial activity and show inhibitory effects on bacterial growth for Gram-positive, Gram-negative bacteria and fungi, suggesting that bombolitin is a potential antimicrobial agent [196]. Osmin, isolated of solitary bee *Osmia rufa*, shows some similarities with the mast cell degranulation (MCD) peptide family. Free acid and C-terminally amidated osmins were chemically synthesized and tested for antimicrobial and haemolytic activities. Antimicrobial and antifungal tests indicated that both peptides were able to inhibit bacterial and fungal growth [197]. Two families of bioactive peptides which belongs to mastoparans (12a and 12b) and chemotactic peptides (5e, 5g and 5f) were purified and characterized from the venom of *Vespa magnifica*. MP-VBs (vespa mastoparan) and VESP-VBs (vespa chemotactic peptide) were purified from the venom of the wasp *Vespa bicolor* Fabricius and demonstrated antimicrobial action [198]. The amphipathic α-helical structure and net positive charge (which permits electrostatic interaction with the negatively charged microbial cell membrane) of mastoparan appear to be critical for MCD activity and because of these structural properties, mastoparans are often highly active against the cell membranes of bacteria, fungi, and erythrocytes, as well as mast cells [199].

6.4. Toxins acting on inflammatory and nociceptive responses

Bee venom has been used in Oriental medicine and evidence from the literature indicates that bee venom plays an anti-inflammatory or anti-nociceptive role against inflammatory reactions associated with arthritis and other inflammatory diseases [200]. Bee venom demonstrated neuroprotective effect against motor neuron cell death and suppresses neuroinflammation-induced disease progression in symptomatic amyotrophic lateral sclerosis (ALS) mice model [200]. Melittin has effects on the secretion of phospholipase A$_2$ and inhibits its enzymatic activity, which is important because phospholipases may release arachidonic acid which is converted into prostaglandins [201]. Have also been reported that melittin decreased the high rate of lethality, attenuated hepatic inflammatory responses, alleviated hepatic pathological injury and inhibited hepatocyte apoptosis. Protective effects were probably carried out through the suppression of NF-jB activation, which inhibited TNF-α liberation. Therefore, melittin may be useful as a potential therapeutic agent for attenuating acute liver injury [202]. In addition of melittin, others agents has shown anti-inflammatory activity. Among them are adolapin and MCDP. Adolapin showed marked anti-inflammatory and anti-nociceptive properties due to inhibition of prostaglandin synthase system [203]. MCDP, isolated of *Apis mellifera* venom, is a strong mediator of mast cell degranulation and releases histamine at low concentrations [204].

6.5. Toxins acting on immunological system

Characterization of the primary structure of allergens is a prerequisite for the design of new diagnostic and therapeutic tools for allergic diseases. Major allergens in bee venom (recognized by IgE in more than 50% of patients) include phospholipase A_2 (PLA$_2$), acid phosphatase, hyaluronidase and allergen C, as well as several proteins of high molecular weights (MWs) [205]. Besides these, Api m 6, was frequently (42%) recognized by IgE from bee venom hypersensitive patients [206]; from wasp venom were purified Vesp c 1 (phospholipase A1) and Vesp c 5 (antigen-5) from *Polistes gallicus*, and Vesp ma 2 and Vesp ma 5 from *Vespa magnifica*, [207-208]. Formulations of poly(lactic-co-glycolic acid) (PLGA) microspheres represent a strategy for replacing immunotherapy in multiple injections of venom. The results obtained with bee venom proteins encapsulated showed that the allergens may still be effective in the induction of an immune response and so may be a new formulation for VIT [209]. Recombinant proteins with immunosuppressive properties have been reported in the literature, such as rVPr1 and rVRr3, identified, cloned and expressed from isolated VPR1 and VPr2 from *Pimpla hypochondriaca* [210]. Chemotactic peptide protonectin 1-6 (ILGTIL-NH2) was detected in the venom of the social wasp *Agelaia pallipes pallipes* [211]. Polybia-MPI and Polybia-CP were isolated from the venom of the social wasp *Polybia paulista* and characterized as chemotactic peptides for PMNL cells [212]. Under the diagnosis, the microarray was reported. Protein chips can be spotted with thousands of proteins or peptides, permitting to analyses the IgE responses against a tremendous variety of allergens. First attempts to microarray with Hymenoptera venom allergens included Api m 1, Api m 2, Ves v 5, Ves g 5 and Pol a 5 in a set-up with 96 recombinant or natural allergen molecules representative of most important allergen sources. The venom allergens from different bee, wasp and ant species can be offered on a single chip, allowing to differentiate the species that has stung based on species-specific markers. The allergen microarray allows the determination and monitoring of allergic patients' IgE reactivity profiles to large numbers of disease-causing allergens by using single measurements and minute amounts of serum [213].

6.6. Toxins with anticancer and cytotoxic activities

Bee venom is the most studied among the arthropods covered in this chapter regarding its anti-cancer activities, due mainly to two substances that have been isolated and characterized: melittin and phospholipase A_2 (PLA$_2$). Melittin and PLA$_2$ are the two major components in the venom of the species *Apis mellifera* [214]. Melittin is inhibitor of calmodulin activity and is an inhibitor of cell growth and clonogenicity of human and murine leukemic cells [215]. Study indicated that key regulators in bee venom-induced apoptosis are Bcl-2 and caspase-3 in human leukemic U937 cells through down-regulation of the ERK and Akt signal pathway [216]. Furthermore recent reports indicate that BV is also able to inhibit tumor growth and exhibit anti-tumor activity *in vitro* and *in vivo* and can be used as a chemotherapeutic agent against malignancy [217]. The adjuvant treatment with PLA$_2$ and phosphatidylinositol-(3,4)-bisphosphate was more effective in the blocking of tumor cell growth [218]. New peptides have been isolated from bee venom and tested in tumor cells, exhibiting promising activities in the treatment of cancer. Lasioglossins isolated from the venom of the bee *Lasioglossum laticeps* exhibited potency to kill various cancer cells *in vitro* [219]. Briefly the bee venom acts inhibiting cell proliferation and promoting cell death by different means: increasing Ca^{2+} influx; inducing cytochrome C release; binding calmodulin; decreasing or increasing the expression of proteins that control cell cycle or activating PLA_2, causing damage to cell membranes interfering in the apoptotic pathway [220]. Among potential anticancer compounds, one of the most studied is mastoparan, peptide isolated from wasp venom that has been reported to induce a potent facilitation of the mitochondrial permeability transition. It should be noted that this recognized action of mastoparan is marked at concentrations <1 μM [221]. Two novel mastoparan peptides, Polybia-MP-II e Polybia-MP-III isolated from venom of the social wasp *Polybia paulista*, exhibited hemolytic activity on erythrocytes [222]. Polybia-MPI, also was purified from the venom of the social wasp *P. paulista*, synthesized and studied its antitumor efficacy and cell selectivity. Results revealed that polybia-MPI exerts cytotoxic and antiproliferative efficacy by pore formation and have relatively lower cytotoxicity to normal cells [223].

6.7. Toxins with insulin releasing activity

Bee venom inhibits insulitis and development of diabetes in non-obese diabetic (NOD) mice. The cumulative incidence of diabetes at 25 weeks of age in control was 58% and NOD mice bee venom treated was 21% [224]. Mastoparan, component of wasp venom, is known to affect phosphoinositide breakdown, calcium influx, exocytosis of hormones and neurotransmitters and stimulate the GTPase activity of guanine nucleotide-binding regulatory proteins [225]. Thus, it is reported in the literature that mastoparan stimulates insulin secretion in human, as well as in rodent. Furthermore, glucose and alpha-ketoisocaproate (alfa-KIC) increase the mastoparan-stimulated insulin secretion [226].

7. Biotechnological and pharmacological applications of ant, centipede and caterpillar venom toxins

Ant, centipede and caterpillar venoms have not been studied so extensively as the venoms of snakes, scorpions and spiders. Ant venoms are rich in the phospholipase A_2 and B, hyaluronidase, and acid and alkaline phosphatase as well as in histamine itself [227]. Centipede venoms have been poorly characterized in the literature. Studies have reported in centipede venoms the presence of esterases, proteinases, alkaline and acid phosphatases, cardiotoxins, histamine, and neurotransmitter releasing compounds in *Scolopendra* genus venoms [228]. Among the most studied caterpillar venoms are *Lonomia obliqua* and *Lonomia achelous* venoms, which cause similar clinical effects [229]. Based on cDNA libraries, was possible to identify several proteins from *L. obliqua*, such as cysteine proteases, group III phospholipase A_2, C-type lectins, lipocalins, in addition to protease inhibitors including serpins, Kazal-type inhibitors, cystatins and trypsin inhibitor-like molecules [230].

7.1. Toxins acting on cardiovascular system

A study showed that the *Lonomia obliqua* caterpillar bristles extract (LOCBE) directly releases kinin from low-molecular weight kininogen, being suggested that kallikrein-kinin system plays a role in the edematogenic and hypotensive effects during *L. obliqua* envenomation [231].

7.2. Toxins acting on hemostasis

There are numerous studies in literature reporting the effects on the hemostatic system of toxins from caterpillars. The effect of a crude extract of spicules from *Lonomia obliqua* caterpillar on hemostasis was found to activate both prothrombin and factor X [232]. Lopap is a prothrombin activator isolated from the bristles of *L. obliqua* caterpillar. Lopap demonstrated ability to induce activation, expression of adhesion molecules and to exert an anti-apoptotic effect on human umbilical vein endothelial cells [233]. Lonofibrase, an α-fibrinogenase from *L. obliqua* was isolated from venomous secretion [234]. Losac, a protein with procoagulant activity, which acts as a growth stimulator and an inhibitor of cellular death for endothelial cells, was purified of the bristle extract of *L. obliqua*. Losac may have biotechnological applications, including the reduction of cell death and consequently increased productivity of animal cell cultures [235]. Lonomin V, serine protease isolated from *Lonomia achelous* caterpillar, inhibited platelet aggregation, probably caused by the degradation of collagen. It is emphasized that Lonomin V shows to be a potentially useful tool for investigating cell-matrix and cell-cell interactions and for the development of antithrombotic agents in terms of their anti-adhesive activities [236]. The venom from the tropical ant, *Pseudomyrmex triplarinus*, inhibited arachidonic acid and induced platelet aggregation, suggesting that venom prevented the action of prostaglandins. The venom was fractionated and factor F (adenosine) with antiplatelet activity were detected [237].

7.3. Toxins with antibiotic activity

Venom alkaloids from *Solenopsis invicta*, fire ant, inhibit the growth of Gram-positive and Gram-negative bacteria and presumably act as a brood antibiotic. Peptides named ponericins were identified from the venom of ant *Pachycondyla goeldii*. Fifteen peptides were classified into three different families according to their primary structure similarities: ponericins G, W, and L. Ponericin G1, G3, G4 and G6 demonstrated antimicrobial activity. Ponericins G share about 60% sequence similarity with cecropins and these have a broad spectrum of activity against bacteria. Peptides family W shares about 70% sequence similarity with Gaegurin 5 (*Rana rugosa*) and melittin (discussed in previous topics). Gaegurin 5 exhibits a broad spectrum of antimicrobial action against bacteria, fungi, and protozoa and has very little hemolytic action. The ponericin L2 from the third family has only an antibacterial action, and shares important sequence similarities with dermaseptin 5, which has strong antimicrobial action against bacteria, yeast, fungi, and protozoa [238]. A cytotoxic peptide from the venom of the ant *Myrmecia pilosula*, Pilosulin 1, was identified as a potential novel antimicrobial peptide sequence. It outlined a potent and broad spectrum antimicrobial activity including standard and multi-drug resistant gram-positive and gram-negative bacteria and

Candida albicans [239]. Two antimicrobial peptides from centipede venoms, scolopin 1 and 2 were identified from venoms of *Scolopendra subspinipes mutilan* [240].

7.4. Toxins acting on inflammatory and nociceptive responses

Venom from the tropical ant *Pseudomyrex triplarinus* relieves pain and inflammation in rheumatoid arthritis [241]. Venom from the *P. triplarinus* contains peptides called myrmexins that relieve pain and inflammation in patients with rheumatoid arthritis and inhibit inflammatory carragenin-induced edema in mice [242].

7.5. Toxins acting on immunological system

The most frequent cause of insect venom allergy in the Southeastern USA is the imported fire ant and the allergens are among the most potent known. Fire ant venom is a potent allergy-inducing agent containing four major allergens, Sol i I, Sol i II, Sol i III and Sol i IV [243-244].

7.6. Toxins with anticancer and cytotoxic activities

Solenopsin A, a primary alkaloid from the fire ant *Solenopsis invicta*, exhibits antiangiogenic activity. Among the results obtained in this study, one of the most interesting was the selective inhibition of Akt by solenopsin *in vitro*, that is of great interest since few Akt inhibitors have been developed, and Akt is a key molecular target in the pharmacological treatment of cancer [245]. Glycosphingolipid 7, identified in the millipede *Parafontaria laminata armigera*, suppressed cell proliferation and this effect was associated with suppression of the activation of FAK (focal adhesion kinase), Erk (extracellular signal-regulated kinase), and Akt in melanoma B16F10 cells. Cells treated with glycosphingolipid 7 reduced the expression of the proteins responsible for the progression of cell cycle, cyclin D1 and CDK4 [246].

7.7. Toxins with insecticides applications

Peptides named ponericins from ant *Pachycondyla goeldii* have a marked action as insecticides. Among the peptides showed insecticidal activity are the ponericins G1, G2 and ponericins belonging to the family W [238].

In Table 1, is presented a summary of the main biotechnological/pharmacological applications of toxins from venomous animals covered in this chapter.

Source	Toxin		Application	Ref.
		Toxins acting on cardiovascular system		
Snakes	*Agkistrodon halys blomhoffii*	NP	Anti-hypertensive agent	[21]
	Bothrops jararaca	BPP	Anti-hypertensive agent (development of captopril and derivatives)	[16]

		NP	Anti-hypertensive agent	[4]
	Bungarus flaviceps	NP	Anti-hypertensive agent	[24]
	Crotalus durissus cascavella	NP	Anti-hypertensive agent	[23]
	Crotalus durissus terrificus	BPP	Anti-hypertensive agent	[17]
	Dendroaspis angusticeps	DNP	Anti-hypertensive agent: natriuretic peptide	[19]
		C10S2C2	Anti-hypertensive drug: L-type Ca^2+channels blocker	[27]
	Dendroaspis jamesoni kaimosae	S4C8	Anti-hypertensive agent: L-type Ca^2+channels blocker	[27]
	Dendroaspis polylepis polylepis	Calciseptine	Anti-hypertensive agent: L-type Ca^2+channels blocker	[25]
		FS2 toxins	Anti-hypertensive agent: L-type Ca^2+channels blocker	[26]
	Micrurus corallinus	NP	Anti-hypertensive agent	[20]
	Pseudocerastes persicus	NP	Anti-hypertensive agent	[22]
	Trimeresurus flavoviridis	NP	Anti-hypertensive agent	[21]
	Trimeresurus stejnegeri	Stejnihagin	Anti-hypertensive agent: L-type Ca^2+channels blocker	[29]
Scorpions	Buthus martensii	BPP	Anti-hypertensive agent	[97]
	Leiurus quinquestriatus	BPP	Anti-hypertensive agent	[98]
	Tityus serrulatus	BPP	Anti-hypertensive agent	[96]
Spiders	Grammostola spatulata	GsMtx-4	Blocks cardiac stretch-activated ion channels and suppresses atrial fibrillation in rabbits	[147]
Toads and Frogs	Atelopus zeteki	Atelopidtoxin	Hypotensive agent and ventricular fibrillation inductor	[167]
	Bufo bufo gargarizans	Bufalin	Na^K+-ATPase inhibitor	[165]
	Pseudophryne coriacea	Semi-purified skin extracts	Hypotensive agent	[168]
	Rana igromaculata	Bradykinin	Hypotensive agent and smooth muscle exciting substance	[11]
	Rana margaratae	Margaratensin	Neurotensin-like peptide	[164]
		Cinobufagin	Na^K+ATPase inhibitor	[165]
	Rana temporaria	Bradykinin	Hypotensive agent and smooth muscle exciting substance	[11]
Bees and Wasps	Apis mellifera	Cardiopep	Beta-adrenergio-like stimulant and anti-arrhythmic agent	[185]

	Vespa basalis	Mastoparan B	Anti-hypertensive agent	[186]
Toxins acting on hemostasis				
Snakes	*Agkistrodon rhodostoma*	Ancrod	Anticoagulant and defibrinogenating agent (Viprinex®)	[34]
	Bothrops alternatus	Bhalternin	Treatment and prevention of thrombotic disorders	[35]
	Bothrops atrox	Batroxobin	Anticoagulant and defibrinogenating agent (Defibrase®)	[33]
		Mixture of a TLE with a thromboplastin-like enzyme	Treatment of hemorrhages (Haemocoagulase®)	[13]
	Bothrops erythromelas	BE-I-PLA²	Antiplatelet agent	[39]
	Bothrops leucurus	BleucMP	Treatment and prevention of cardiovascular disorders and strokes	[36]
		Leucurogin	Antiplatelet agent	[32]
	Echis carinatus	Echistatin	Antiplatelet agent	[30]
	Sisturus miliaris barbouri	Barbourin	Antiplatelet agent	[31]
	Trimeresurus malabaricus	Trimarin	Treatment and prevention of thrombotic disorders	[38]
	Vipera lebetina	VLH2	Treatment and prevention of thrombotic disorders	[37]
Spiders	*Loxosceles.*	Phospholipase-D	Platelet aggregation inductor	[149]
Toads and Frogs	*Bombina maxima*	Bm-ANXA2	Antiplatelet agent	[169]
Bees and Wasps	*Bombus hypocrita sapporoensis*	Bs-VSP	Prothrombin activator, thrombin-like protease and a plasmin-like protease agent	[190]
	Bombus ignites	Bi-VSP	Prothrombin activator, thrombin-like protease and a plasmin-like protease agent	[188]
		Bi-KTI	Plasmin inhibitor agent	[187]
	Bombus terrestris	Bt-VSP	Prothrombin activator, thrombin-like protease and a plasmin-like protease agent	[189]
	Vespa orientalis	Protease I	Anticoagulant agent	[192]
	Vespa magnifica	Magnifin	Inductor platelet aggregation agent	[193]
		Magnvesin	Anticoagulant agent	[191]
Ants, Centipedes and Caterpillars	*Lonomia achelous*	Lonomin V	Inhibitor platelet aggregation agent	[236]
		Lopap	Prothrombin activator agent	[233]
	Lonomia obliqua	Lonofibrase	Fibrinogenolytic and fibrinolytic agent Agent	[234]
		Losac	Procoagulant agent	[235]
Toxins with antibiotic activity				
Snakes	*Bothrops alternatus*	Balt-LAAO-I	Anti-bacterial agent	[42]

	Bothrops asper	Myotoxin II	Anti-bacterial agent	[50]
	Bothrops jararaca	LAAO	Antiparasitic agent	[48]
	Bothrops marajoensis	BmarLAAO	Anti-bacterial, antifungal and antiparasitic agent	[47]
	Bothrops neuwiedi	Neuwiedase	Antiparasitic agent	[55]
	Bothrops pirajai	BpirLAAO-I	Anti-bacterial and antiparasitic agent	[44]
	Bungarus fasciatus	BFPA	Anti-bacterial agent	[53]
	Crotalus durissus cascavella	Casca LAO	Anti-bacterial agent	[45]
	Crotalus durissus terrificus	Crotoxin	Antiviral agent	[56]
		PLA$_2$-CB	Antiviral agent	[56]
		PLA$_2$-IC	Antiviral agent	[56]]
	Echis carinatus	EcTx-I	Anti-bacterial agent	[51]
	Naja atra	Vgf-1	Anti-bacterial agent	[54]
	Naja naja oxiana	LAAO	Anti-bacterial agent	[46]
	Porthidium nasutum	PnPLA$_2$	Anti-bacterial agent	[52]
	Trimeresurus jerdonii	TJ-LAO	Anti-bacterial agent	[41]
	Trimeresurus mucrosquamatus	TM-LAO	Anti-bacterial agent	[43]
Scorpions	Hadrurus aztecus	Hadrurin	Anti-bacterial agent	[110]
	Leiurus quinquestriatus	Defensin	Anti-bacterial agent	[102]
	Opistophthalmus carinatus	Opistoporin I/II	Anti-bacterial and antifungal agent	[106]
	Pandinus imperator	Pandinin I/II	Antimicrobial agent	[101]
		Scorpine	Anti-bacterial and antiparasitic agent	[104]
	Parabuthus schlechteri	Cationic amphipatic peptide	Antimicrobial agent	[109]
	Tityus discrepans	Bactridines	Anti-bacterial agent	[111]
Spiders	Lycosa carolinensis	Lycotoxins I/II	Antimicrobial agent	[150]
Toads and Frogs	Bufo bufo gargarizans	6-methyl-spinaceamine	Anti-bacterial agent	[172]
		Bufalin	Antiviral agent	[173]
		Cinobufagin	Antiviral agent	[173]
	Bufo rubescens	Telocinobufagin	Anti-bacterial agent	[170]
		Marinobufagin	Anti-bacterial agent	[170]
	Leptodactylus pentadactylus	Apinaceamine	Anti-bacterial agent	[172]
	Leptodactylus syphax	SPXs	Anti-bacterial agent	[171]
	Rhinella jimi	Telocinobufagin	Antiparasitic agent	[174]

		Hellebrigenin	Antiparasitic agent	[174]
Bees and Wasps	*Apis mellifera*	Melittin	Anti-bacterial agent	[195]
	Bombus ignites	Bi-Bombolitin	Anti-bacterial and antifungal agent	[196]
	Osmia rufa	Osmin	Anti-bacterial and antifungal agent	[197]
	Vespa bicolor	MP-VB1	Anti-bacterial and antifungal agent	[198]
		VESP-VB1	Anti-bacterial and antifungal agent	[198]
Ants, Centipedes and Caterpillars	*Myrmecia pilosula*	Pilosulin 1	Anti-bacterial and antifungal agent	[239]
	Scolopendra	Scolopin 1	Anti-bacterial and antifungal agent	[240]
	subspinipes mutilan	Scolopin 2	Anti-bacterial and antifungal agent	[240]
Toxins acting on inflammatory and nociceptive responses				
Snakes	*Crotalus durissus terrificus*	Crotamine	Antinociceptive agent	[63]
		Crotoxin	Antinociceptive agent	[64]
		Hyal	Anti-edematogenic agent	[59]
	Lachesis muta	βPLI	Phospholipase inhibitor	[58]
	Naja atra	Cobrotoxin	Antinociceptive agent	[65]
	Ophiophagus hannah	Hannalgesin	Antinociceptive agent	[66]
Scorpions	*Buthus martensii*	BmKIT2	Antinociceptive agent	[115]
		J123 peptide	K+ channel blocker	[112]
Spiders	*Loxosceles laeta*	SMase D	Pro-inflammatory agent	[152]
	Loxosceles reclusa	Phospholipase D	Pro-inflammatory agent	[152]
	Psalmopoeus cambridgei	Psalmotoxin 1	Antinociceptive and anti-inflammatory agent	[151]
Toads and Frogs	*Epipedobates tricolor*	Epibatidine	Antinociceptive agent	[175]
	Phyllomedusa sp	Deltorphins	Opioid analgesic agents	[176]
		Dermorphins	Opioid analgesic agents	[61]
Bees and Wasps	*Apis mellifera*	Melittin	Anti-inflammatory agent	[202]
		MCDP	Anti-inflammatory agent	[204]
Ants, Centipedes and Caterpillars	*Pseudomyrex triplarinus*	Myrmexins	Anti-inflammatory agent	[242]
Toxins acting on immunological system				
Snakes	*Crotalus durissus terrificus*	Crotapotin	Immunossupressive agent	[69]
		Crotoxin	Immunossupressive agent	[68]
	Ophiophagus hannah	OVF	Complement system activator agent	[71]
Scorpions	*Androctonus mauretanicus*	Kaliotoxin	Ca^{2+} activated K+ channel	[121]
	Centruroides limbatus	Hongotoxin	K+ channel blocker	[123]
	Centruroides margaritatus	Margatoxin	Immunosuppressive agent	[120]
	Centruroides noxius	Noxiustoxin	K+ channel blocker	[124]

	Leiurus quinquestriatus	Agitoxin I/II/III	K+ channel blocker	[122]
	Orthochirus scrobiculosus	OSK1	Immunosuppressive agent	[119]
	Pandinus imperator	Pi1	K+ channel blocker	[125]
Spiders	*Loxosceles laeta*	SMase D	Antiserum	[155]
	Loxosceles reclusa	SMase D	Antiserum	[155]
Bees and Wasps	*Agelaia pallipes pallipes*	Protonectin 1-6	Chemotactic agent	[211]
	Apis mellifera	Api m 1	Allergen	[213]
		Api m 2	Allergen	[213]
		Api m 6	Allergen	[206]
	Pimpla hypochondriaca	rVPr1	Immunosuppressive agent	[210]
		rVPr3	Immunosuppressive agent	[210]
	Polistes annularis	Pol a 5	Allergen	[213]
	Polistes gallicus	Vesp c 1 (phospholipase A1)	Allergen	[207-208]
		Vesp c 5 (antigen-5)	Allergen	[207-208]
	Polybia paulista	Polybia-MPI	Chemotactic agent	[212]
		Polybia-CP	Chemotactic agent	[212]
	Vespa magnifica	Vesp ma 2	Allergen agent	[207-208]
		Vesp ma 5	Allergen	[207-208]
	Vespula germanica	Ves g 5	Allergen	[213]
	Vespula vulgaris	Ves v 5	Allergen	[213]
Ants, Centipedes and Caterpillars	*Solenopsis invicta*	Sol i I	Allergen	[244]
		Sol i II	Allergen	[243]
		Sol i III	Allergen	[243]
		Sol i IV	Allergen	[243]
		Toxins with anticancer and cytotoxic activity		
Snakes	*Agkistrodon acutus*	Accutin	Anticancer agent: disintegrin	[73]
		ACTX-6	Anticancer agent: L-amino acid oxidase	[82]
	Agkistrodon contortrix	Contortrostatin	Anticancer agent: disintegrin	[75]
	Agkistrodon halys brevicaudus	Salmosin	Anticancer agent: disintegrin	[74]
	Bothrops brazili	sPLA$_2$	Anticancer agent	[86]
	Bothrops jararacussu	BJcuL	Anticancer agent	[89]
	Bothrops leucurus	Bl-LAAO	Anticancer agent	[84]

		Metalloproteinase	Anticancer agent	[90]
		Lectin	Anticancer agent	[91]
	Bothrops neuwiedii	sPLA$_2$	Anticancer agent	[85]
	Calloselasma rhodostoma	Rhodostomin	Anticancer agent: disintegrin	[78]
	Crotalus atrox	Crotatroxin	Anticancer agent: disintegrin	[77]
	Naja naja	Disintegrin	Anticancer agent	[79]
	Naja naja naja	sPLA$_2$	Anticancer agent	[87]
	Ophiophagus hannah	LAAO	Anticancer agent	[81]
	Trimeresurus flavoviridis	OHAP-1	Anticancer agent: L-amino acid oxidase	[83]
	Trimeresurus jerdonii	Jerdonin	Anticancer agent: disintegrin	[76]
Scorpions	*Heterometrus bengalensis*	Bengalin	Anticancer agent	[116]
	Leiurus quinquestriatus	Chlorotoxin	Anticancer agent	[126]
		rBmK CTa	Anticancer agent	[130]
	Tityus discrepans	Neopladine 1 and 2	Anticancer agent	[131]
Spiders	*Acanthoscurria gomesiana*	Gomesin	Cytotoxic and anticancer agent	[157]
	Psalmopoeus cambridgei	Psalmotoxin 1	Anticancer agent	[156]
Toads and Frogs	*Bombina variegata pachypus*	Cutaneous venom	Anticancer agent	[180]
	Bufo bufo gargarizans	Bufalin	Anticancer agent	[178]
		Cinobufagin	Anticancer agent	[178]
	Formosan Ch'an Su	Bufotalin	Anticancer agent	[179]
	Rana ridibunda	Brevinin-2R	Anticancer agent	[181]
	Venenum Bufonis	CBG	Anticancer and immunotherapeutic agent to treat immune-mediated diseases	[177]
Bees and Wasps	*Lasioglossum laticeps*	Lasioglossins	Anticancer agent	[219]
	Polybia paulista	Polybia-MPI	Cytotoxic and antiproliferative agent	[223]
		Polybia-MP-II	Cytotoxic agent (hemolytic activity on erythrocytes)	[222]
		Polybia-MP-III	Cytotoxic agent (hemolytic activity on erythrocytes)	[222]
Ants, Centipedes and Caterpillars	*Parafontaria laminata armigera*	Glycosphingolipid 7	Anticancer agent	[246]
	Solenopsis invicta	Solenopsin A	Anticancer agent	[245]

Toxins with insulin releasing activity

Toads and Frogs	*Agalychnis litodryas*	Peptides from skin secretion	Insulin-releasing activity	[182]
Bees and Wasps	Wasp venom	Mastoparan	Stimulator of insulin secretion agent	[226]
		Toxins with insecticides applications		
Scorpions	*Androctonus australis*	AaH IT1	Anti-insect agent	[142]
	Buthotus judaicu	Bjα IT	Anti-insect agent	[137]
	Buthus martensii	BmKM1	Anti-insect agent	[140]
	Buthus martensii	Bm 32/33	Anti-insect agent	[144]
	Buthus occitanus	Bot IT1	Anti-insect agent	[135]
	Buthus occitanus mardochei	Bom III/IV	Anti-insect agent	[139]
	Leiurus quinquestriatus	Lqhα IT	Anti-insect agent	[134]
	Leiurus quinquestriatus hebraeus	Lqh III/ VI/ VII	Anti-insect agent	[141]
	Odonthobuthus doriae	OD1	Anti-insect agent	[137]
Spiders	*Loxosceles arizonica*	SMase D	Anti-insect agent	[161]
	Loxosceles intermedia	LiTxx1/ LiTxx2/ LiTxx3	Anti-insect agent	[158]
	Paracoelotes luctuosus	δ-PaluIT1/ δ-PaluIT2	Anti-insect agent	[162]
	Phoneutria nigriventer	Tx4(6-1)	Anti-insect agent	[160]
Ants, Centipedes and Caterpillars		Ponericins G1	Insecticide Agent	[238]
	Pachycondyla goeldii	Ponericins G2	Insecticide Agent	[238]
		Ponericins family W	Insecticide Agent	[238]

Table 1. Summary of the main biotechnological/pharmacological applications of toxins from venomous animals.

8. Conclusion

The biodiversity of venoms and toxins made it a unique source of leads and structural templates from which new therapeutic agents may be developed. Such richness can be useful to biotechnology and/or pharmacology in many ways, with the prospection of new toxins in this field. Venoms of several animal species such as snakes, scorpions, toads, frogs and their active components have shown potential biotechnological applications. Recently, using molecular biology techniques and advanced methods of fractionation, researchers have obtained different native and/or recombinant toxins and enough material to afford deeper insight into the molecular action of these toxins. The mechanistic elucidation of toxins as well as their use as drugs will depend on insight into toxin biochemical classification, structure/conformation determination and elucidation of toxin biological activities based on their molecular organization, in addition to their mechanism of action upon different cell models as well as their cellular receptors. Furthermore, expansions in the fields of chemistry and biology have guided new drug discovery strategies to maximize the identification of biotechnological relevant toxins. In fact, with so much diversity in the terrestrial fauna to be explored in the future, is extremely important providing a further stimulus to the preservation of the precious ecosystem in order to develop the researches focusing on identify and isolate new molecules with importance in biotechnology or pharmacology.

Acknowledgements

Our research on this field is supported by Fundação de Amparo à Pesquisa do Estado do Rio Grande do Norte (FAPERN), Coordenação de Aperfeiçoamento de Pessoal de Nível Superior (CAPES) and Conselho Nacional de Desenvolvimento Científico e Tecnológico (CNPq).

Author details

Matheus F. Fernandes-Pedrosa*, Juliana Félix-Silva and Yamara A. S. Menezes

*Address all correspondence to: mpedrosa@ufrnet.br

Universidade Federal do Rio Grande do Norte, Brazil

References

[1] Gomes, A., Bhattacharjee, P., Mishra, R., Biswas, A. K., Dasgupta, S. C., Giri, B., Debnath, A., Gupta, S. D., & Das, T. (2010). Anticancer potential of animal venoms and toxins. *Indian Journal of Experimental Biology*, 48(2), 93-103.

[2] Braud, S., Bon, C., & Wisner, A. (2000). Snake venom proteins acting on hemostasis. *Biochimie*, 82(9-10), 851-9.

[3] Kang, T. S., Georgieva, D., Genov, N., Murakami, M. T., Sinha, M., Kumar, R. P., Kaur, P., Kumar, S., Dey, S., Sharma, S., Vrielink, A., Betzel, C., Takeda, S., Arni, R. K., Singh, T. P., & Kini, R. M. (2011). Enzymatic toxins from snake venom: Structural characterization and mechanism of catalysis. *FEBS Journal*, 278(23), 4544-4576.

[4] Murayama, N., Hayashi, M. A. F., Ohi, H., Ferreira, L. A. F., Hermann, V. V., Saito, H., Fujita, Y., Higuchi, S., Fernandes, B. L., Yamane, T., & Camargo, A. C. M. (1997). Cloning and sequence analysis of a Bothrops jararaca cDNA encoding a precursor of seven bradykinin-potentiating peptides and a C-type natriuretic peptide. *Proceedings of the National Academy of Sciences of the United States of America*, 94(4), 1189-1193.

[5]   Quintero-Hernández, V., Ortiz, E., Rendón-Anaya, M., Schwartz, E. F., Becerril, B., Corzo, G., & Possani, L. D. (2011). Scorpion and spider venom peptides: Gene cloning and peptide expression. *Toxicon*, 58(8), 644-663.

[6]   Verano-Braga, T., Rocha-Resende, C., Silva, D. M., Ianzer, D., Martin-Euaclaire, M. F., Bougis, P. E., De Lima, ME, Santos, R. A. S., & Pimenta, A. M. C. (2008). Tityus serrulatus Hypotensins: A new family of peptides from scorpion venom. *Biochemical and Biophysical Research Communications*, 371(3), 515-520.

[7]   Almeida, D. D., Scortecci, K. C., Kobashi, L. S., Agnez-Lima, L. F., Medeiros, S. R. B., Silva-Junior, A. A., Junqueira-de-Azevedo, I. L. M., & Fernandes-Pedrosa, M. F. (2012). Profiling the resting venom gland of the scorpion Tityus stigmurus through a transcriptomic survey. *BMC Genomics*, 13, 362.

[8]   Petricevich, V. L. (2010). Scorpion venom and the inflammatory response. *Mediators of Inflammation*, 903295.

[9]   Tambourgi, D. V., Pedrosa, M. F. F., Van Den, Berg. C. W., Gonçalves-de-Andrade, R. M., Ferracini, M., Paixão-Cavalcante, D., Morgan, B. P., & Rushmere, N. K. (2004). Molecular cloning, expression, function and immunoreactivities of members of a gene family of sphingomyelinases from Loxosceles venom glands. *Molecular Immunology*, 41(8), 831-840.

[10] Senff-Ribeiro, A., Henrique, da., Silva, P., Chaim, O. M., Gremski, L. H., Paludo, K. S., Bertoni da. Silveira, R., Gremski, W., Mangili, O. C., & Veiga, S. S. (2008). Biotechnological applications of brown spider (Loxosceles genus) venom toxins. *Biotechnology Advances*, 26(3), 210-218.

[11] Habermehl, GG. (1995). Antimicrobial activity of amphibian venoms. *Studies in Natural Products Chemistry, Part C*, 327-339.

[12] Habermann, E. (1972). Bee and wasp venoms. *Science*, 177(4046), 314-322.

[13] Sajevic, T., Leonardi, A., & Križaj, I. (2011). Haemostatically active proteins in snake venoms. *Toxicon*, 57(5), 627-645.

[14] Koh, C. Y., & Kini, R. M. (2012). From snake venom toxins to therapeutics- cardiovascular examples. *Toxicon*, 59(4), 497-506.

[15] Hodgson, W. C., & Isbister, G. K. (2009). The application of toxins and venoms to cardiovascular drug discovery. *Current Opinion in Pharmacology*, 9(2), 173-176.

[16] Ferreira, S. H. (1965). A bradykinin-potentiating factor (BPF) present in the venom of Bothrops jararaca. *British Journal of Pharmacology and Chemotherapy*, 24(1), 163-169.

[17] Barreto, S. A., Chaguri, L. C. A. G., Prezoto, B. C., & Lebrun, I. (2012). Characterization of two vasoactive peptides isolated from the plasma of the snake Crotalus durissus terrificus. Biomedicine and Pharmacotherapy , 66(4), 256-265.

[18] Vink, S., Jin, A. H., Poth, K. J., Head, G. A., & Alewood, P. F. (2012). Natriuretic peptide drug leads from snake venom. *Toxicon*, 59(4), 434-445.

[19] Schweitz, H., Vigne, P., Moinier, D., Frelin, C., & Lazdunski, M. (1992). A new member of the natriuretic peptide family is present in the venom of the green mamba (Dendroaspis angusticeps). Journal of Biological Chemistry Jul 15; , 267(20), 13928-13932.

[20] Ho, P. L., Soares, M. B., Maack, T., Gimenez, I., Puorto, G., Furtado, M. D. F. D., & Raw, I. (1997). Cloning of an unusual natriuretic peptide from the South American coral snake Micrurus corallinus. *European Journal of Biochemistry*, 250(1), 144-149.

[21] Higuchi, S., Murayama, N., Saguchi, K., Ohi, H., Fujita, Y., Camargo, A. C. M., Ogawa, T., Deshimaru, M., & Ohno, M. (1999). Bradykinin-potentiating peptides and C-type natriuretic peptides from snake venom. *Immunopharmacology*, 44(1-2), 129-135.

[22] Amininasab, M., Elmi, M. M., Endlich, N., Endlich, K., Parekh, N., Naderi-Manesh, H., Schaller, J., Mostafavi, H., Sattler, M., Sarbolouki, M. N., & Muhle-Goll, C. (2004). Functional and structural characterization of a novel member of the natriuretic family of peptides from the venom of Pseudocerastes persicus. *FEBS Letters*, 557(1-3), 104-108.

[23] Evangelista, J. S. A. M., Martins, A. M. C., Nascimento, N. R. F., Sousa, C. M., Alves, R. S., Toyama, D. O., Toyama, M. H., Evangelista, J. J. F., Menezes, D. B., Fonteles, M. C., Moraes, M. E. A., & Monteiro, H. S. A. (2008). Renal and vascular effects of the natriuretic peptide isolated from Crotalus durissus cascavella venom. *Toxicon*, 52(7), 737-744.

[24] Siang, AS, Doley, R., Vonk, F. J., & Kini, R. M. (2010). Transcriptomic analysis of the venom gland of the red-headed krait (Bungarus flaviceps) using expressed sequence tags. *BMC Molecular Biology*, 11.

[25] De Weille, J. R., Schweitz, H., Maes, P., Tartar, A., & Lazdunski, M. (1991). Calciseptine, a peptide isolated from black mamba venom, is a specific blocker of the L-type calcium channel. *Proceedings of the National Academy of Sciences of the United States of America*, 88(6), 2437-2440.

[26] Yasuda, O., Morimoto, S., Jiang, B., Kuroda, H., Kimura, T., Sakakibara, S., Fukuo, K., Chen, S., Tamatani, M., & Ogihara, T. (1994). FS2. A mamba venom toxin, is a specific blocker of the L-type calcium channels. *Artery DCOM- 19961204*, 21(5), 287-302.

[27] Joubert, F. J., & Taljaard, N. (1980). The complete primary structures of two reduced and S-carboxymethylated Angusticeps-type toxins from Dendroaspis angusticeps (green mamba) venom. *BBA- Protein Structure*, 623(2), 449-456.

[28] Joubert, F. J., & Taljaard, N. (1980). The primary structure of a short neurotoxin homologue (S4C8) from Dendroaspis jamesoni kaimosae (Jameson's mamba) venom. *International Journal of Biochemistry*, 12(4), 567-574.

[29] Zhang, P., Shi, J., Shen, B., Li, X., Gao, Y., Zhu, Z., Zhu, Z., Ji, Y., Teng, M., & Niu, L. (2009). Stejnihagin, a novel snake metalloproteinase from Trimeresurus stejnegeri venom, inhibited L-type Ca^{2+} channels. *Toxicon*, 53(2), 309-315.

[30] Bilgrami, S., Tomar, S., Yadav, S., Kaur, P., Kumar, J., Jabeen, T., Sharma, S., & Singh, T. P. (2004). Crystal structure of Schistatin, a disintegrin homodimer from Saw-scaled Viper (Echis carinatus) at 2.5 Å Resolution. *Journal of Molecular Biology*, 341(3), 829-837.

[31] Scarborough, R. M., Rose, J. W., Hsu, MA, Phillips, D. R., Fried, V. A., Campbell, A. M., Nannizzi, L., & Charo, I. F. (1991). Barbourin: a GPIIb-IIIa-specific integrin antagonist from the venom of Sistrurus m. barbouri. *Journal of Biological Chemistry*, 266(15), 9359-9362.

[32] Higuchi, D. A., Almeida, M. C., Barros, C. C., Sanchez, E. F., Pesquero, P. R., Lang, E. A. S., Samaan, M., Araujo, R. C., Pesquero, J. B., & Pesquero, J. L. (2011). Leucurogin, a new recombinant disintegrin cloned from Bothrops leucurus (white-tailed-jararaca) with potent activity upon platelet aggregation and tumor growth. *Toxicon*, 58(1), 123-129.

[33] Stocker, K., & Barlow, G. H. (1976). The coagulant enzyme from Bothrops atrox venom (batroxobin). *Methods in Enzymology*, 45-214.

[34] Nolan, C., Hall, L. S., & Barlow, G. H. (1976). Ancrod, the coagulating enzyme from Malayan pit viper (Agkistrodon rhodostoma) venom. *Methods in Enzymology*, 45-205.

[35] Costa, J. O., Fonseca, K. C., Mamede, C. C. N., Beletti, M. E., Santos-Filho, N. A., Soares, A. M., Arantes, E. C., Hirayama, S. N. S., Selistre-de-Araújo, H. S., Fonseca, F., Henrique-Silva, F., Penha-Silva, N., & Oliveira, F. (2010). Bhalternin: functional and structural characterization of a new thrombin-like enzyme from Bothrops alternatus snake venom. *Toxicon*, 55(7), 1365-1377.

[36] Gomes, M. S. R., Queiroz, M. R., Mamede, C. C. N., Mendes, MM, Hamaguchi, A., Homsi-Brandeburgo, M. I., Sousa, M. V., Aquino, E. N., Castro, Oliveira. F., & Rodrigues, V. M. (2011). Purification and functional characterization of a new metalloproteinase (BleucMP) from Bothrops leucurus snake venom. *Comparative Biochemistry and Physiology, Part C: Toxicology and Pharmacology*, 153(3), 290-300.

[37] Hamza, L., Gargioli, C., Castelli, S., Rufini, S., & Laraba-Djebari, F. (2010). Purification and characterization of a fibrinogenolytic and hemorrhagic metalloproteinase isolated from Vipera lebetina venom. *Biochimie*, 92(7), 797-805.

[38] Kumar, R. V., Gowda, C. D. R., Shivaprasad, H. V., Siddesha, J. M., Sharath, B. K., & Vishwanath, B. S. (2010). Purification and characterization of 'Trimarin' a hemorrhagic metalloprotease with factor Xa-like activity, from Trimeresurus malabaricus snake venom. *Thrombosis Research*, 126(5), e356-e364.

[39] Modesto, J. C. A., Spencer, P. J., Fritzen, M., Valença, R. C., Oliva, M. L. V., Silva, M. B., Chudzinski-Tavassi, A. M., & Guarnieri, M. C. (2006). BE-I-PLA$_2$, a novel acidic phospholipase A$_2$ from Bothrops erythromelas venom: isolation, cloning and characterization as potent anti-platelet and inductor of prostaglandin I$_2$ release by endothelial cells. *Biochemical Pharmacology*, 72(3), 377-384.

[40] Chopra, I., Hesse, L., & O'Neill, A. J. (2002). Exploiting current understanding of antibiotic action for discovery of new drugs. Symposium Series Society for Applied Microbiology; , 31, 4S-15S.

[41] Lu, Q. M., Wei, Q., Jin, Y., Wei, J. F., Wang, W. Y., & Xiong, Y. L. (2002). L-amino acid oxidase from Trimeresurus jerdonii snake venom: purification, characterization, platelet aggregation-inducing and antibacterial effects. *Journal of Natural Toxins*, 11(4), 345-352.

[42] Stábeli, R. G., Marcussi, S., Carlos, G. B., Pietro, R. C. L. R., Selistre-de-Araújo, H. S., Giglio, J. R., Oliveira, E. B., & Soares, A. M. (2004). Platelet aggregation and antibacterial effects of an L-amino acid oxidase purified from Bothrops alternatus snake venom. *Bioorganic and Medicinal Chemistry*, 12(11), 2881-2886.

[43] Wei, J. F., Wei, Q., Lu, Q. M., Tai, H., Jin, Y., Wang, W. Y., & Xiong, Y. L. (2003). Purification, characterization and biological activity of an L-amino acid oxidase from Trimeresurus mucrosquamatus venom. *Acta Biochimica et Biophysica Sinica*, 35(3), 219-224.

[44] Izidoro, L. F. M., Ribeiro, M. C., Souza, G. R. L., Sant'Ana, C. D., Hamaguchi, A., Homsi-Brandeburgo, M. I., Goulart, L. R., Beleboni, R. O., Nomizo, A., Sampaio, S. V., Soares, A. M., & Rodrigues, V. M. (2006). Biochemical and functional characterization of an L-amino acid oxidase isolated from Bothrops pirajai snake venom. *Bioorganic and Medicinal Chemistry*, 14(20), 7034-7043.

[45] Toyama, M. H., Toyama, D. D. O., Passero, L. F. D., Laurenti, M. D., Corbett, C. E., Tomokane, T. Y., Fonseca, F. V., Antunes, E., Joazeiro, P. P., Beriam, L. O. S., Martins, M. A. C., Monteiro, H. S. A., & Fonteles, M. C. (2006). Isolation of a new L-amino acid oxidase from Crotalus durissus cascavella venom. *Toxicon*, 47(1), 47-57.

[46] Samel, M., Tõnismägi, K., Rönnholm, G., Vija, H., Siigur, J., Kalkkinen, N., & Siigur, E. (2008). L-amino acid oxidase from Naja naja oxiana venom. *Comparative Biochemistry and Physiology Part B: Biochemistry and Molecular Biology*, 149(4), 572-580.

[47] Torres, A. F. C., Dantas, R. T., Toyama, M. H., Diz-Filho, E., Zara, F. J., Queiroz, M. G. R., Nogueira, N. A. P., Oliveira, M. R., Toyama, D. O., Monteiro, H. S. A., & Martins, A. M. C. (2010). Antibacterial and antiparasitic effects of Bothrops marajoensis venom and its fractions: phospholipase A2 and L-amino acid oxidase. *Toxicon*, 55(4), 795-804.

[48] Deolindo, P., Teixeira-Ferreira, A. S., Da Matta, R. A., & Alves, E. W. (2010). L-Amino acid oxidase activity present in fractions of Bothrops jararaca venom is responsible for the induction of programmed cell death in Trypanosoma cruzi. *Toxicon*, 56(6), 944-955.

[49] Arni, R. K., & Ward, R. J. (1996). Phospholipase A2- a structural review. *Toxicon*, 34(8), 827-841.

[50] Santamaría, C., Larios, S., Angulo, Y., Pizarro-Cerda, J., Gorvel-P, J., Moreno, E., & Lomonte, B. (2005). Antimicrobial activity of myotoxic phospholipases A_2 from crotalid snake venoms and synthetic peptide variants derived from their C-terminal region. *Toxicon*, 45(7), 807-815.

[51] Samy, R. P., Gopalakrishnakone, P., Bow, H., Puspharaj, P. N., & Chow, V. T. K. (2010). Identification and characterization of a phospholipase A2 from the venom of the Saw-scaled viper: novel bactericidal and membrane damaging activities. *Biochimie*, 92(12), 1854-1866.

[52] Vargas, L. J., Londoño, M., Quintana, J. C., Rua, C., Segura, C., Lomonte, B., & Núñez, V. (2012). An acidic phospholipase A_2 with antibacterial activity from Porthidium nasutum snake venom. *Comparative Biochemistry and Physiology Part B: Biochemistry and Molecular Biology*, 161(4), 341-347.

[53] Xu, C., Ma, D., Yu, H., Li, Z., Liang, J., Lin, G., Zhang, Y., & Lai, R. (2007). A bactericidal homodimeric phospholipases A_2 from Bungarus fasciatus venom. *Peptides*, 28(5), 969-973.

[54] Xie, J. P., Yue, J., Xiong, Y. L., Wang, W. Y., Yu, S. Q., & Wang, H. H. (2003). In vitro activities of small peptides from snake venom against clinical isolates of drug-resistant Mycobacterium tuberculosis. *International Journal of Antimicrobial Agents*, 22(2), 172-174.

[55] Bastos, L. M., Júnior, R. J. O., Silva, D. A. O., Mineo, J. R., Vieira, C. U., Teixeira, D. N. S., Homsi-Brandeburgo, M. I., Rodrigues, V. M., & Hamaguchi, A. (2008). Toxoplasma gondii: Effects of neuwiedase, a metalloproteinase from Bothrops neuwiedi snake venom, on the invasion and replication of human fibroblasts in vitro. *Experimental Parasitology*, 120(4), 391-396.

[56] Muller, V. D. M., Russo, R. R., Oliveira, Cintra. A. C., Sartim, M. A., Alves-Paiva, R.d. M., Figueiredo, L. T. M., Sampaio, S. V., & Aquino, V. H. (2012). Crotoxin and phospholipases A2 from Crotalus durissus terrificus showed antiviral activity against dengue and yellow fever viruses. *Toxicon*, 59(4), 507-515.

[57] Kini, R. M. (2003). Excitement ahead: structure, function and mechanism of snake venom phospholipase A2 enzymes. *Toxicon*, 42(8), 827-840.

[58] Lima, R. M., Estevão-Costa, M. I., Junqueira-de-Azevedo, I. L. M., Lee, Ho. P., Vasconcelos, Diniz. M. R., & Fortes-Dias, C. L. (2011). Phospholipase A_2 inhibitors (βPLIs) are encoded in the venom glands of Lachesis muta (Crotalinae, Viperidae) snakes. *Toxicon*, 57(1), 172-175.

[59] Bordon, K. C. F., Perino, M. G., Giglio, J. R., & Arantes, E. C. (2012). 198. Isolation, enzymatic characterization and action as spreading factor of a hyaluronidase from Crotalus durissus terrificus snake venom. *Toxicon*, 60(2), 197.

[60] Garcia, F., Toyama, M. H., Castro, F. R., Proença, P. L., Marangoni, S., & Santos, L. M. B. (2003). Crotapotin induced modification of T lymphocyte proliferative response through interference with PGE$_2$ synthesis. *Toxicon*, 42(4), 433-437.

[61] Rajendra, W., Armugam, A., & Jeyaseelan, K. (2004). Toxins in anti-nociception and anti-inflammation. *Toxicon*, 44(1), 1-17.

[62] Giorgi, R., Bernardi, MM, & Cury, Y. (1993). Analgesic effect evoked by low molecular weight substances extracted from Crotalus durissus terrificus venom. *Toxicon*, 31(10), 1257-1265.

[63] Mancin, A. C., Soares, A. M., Andrião-Escarso, S. H., Faça, V. M., Greene, L. J., Zuccolotto, S., Pelá, I. R., & Giglio, J. R. (1998). The analgesic activity of crotamine, a neurotoxin from Crotalus durissus terrificus (South American rattlesnake) venom: A biochemical and pharmacological study. *Toxicon*, 36(12), 1927-1937.

[64] Zhang, H. L., Han, R., Chen, Z. X., Chen, B. W., Gu, Z. L., Reid, P. F., Raymond, L. N., & Qin, Z. H. (2006). Opiate and acetylcholine-independent analgesic actions of crotoxin isolated from Crotalus durissus terrificus venom. *Toxicon*, 48(2), 175-182.

[65] Chen, R., & Robinson, S. E. (1992). The effect of cobrotoxin on cholinergic neurons in the mouse. *Life Sciences*, 51(13), 1013-1019.

[66] Pu, X. C., Wong, P. T. H., & Gopalakrishnakone, P. (1995). A novel analgesic toxin (Hannalgesin) from the venom of king cobra (Ophiophagus hannah). *Toxicon*, 33(11), 1425-1431.

[67] Mirshafiey, A. (2007). Venom therapy in multiple sclerosis. *Neuropharmacology*, 53(3), 353-361.

[68] Rangel-Santos, A., Lima, C., Lopes-Ferreira, M., & Cardoso, D. F. (2004). Immunosuppresive role of principal toxin (crotoxin) of Crotalus durissus terrificus venom. *Toxicon*, 44(6), 609-616.

[69] Castro, F. R., Farias, AS, Proença, P. L. F., De La Hoz, C., Langone, F., Oliveira, E. C., Toyama, M. H., Marangoni, S., & Santos, L. M. B. (2007). The effect of treatment with crotapotin on the evolution of experimental autoimmune neuritis induced in Lewis rats. *Toxicon*, 49(3), 299-305.

[70] Vogel, C. W., & Fritzinger, D. C. (2010). Cobra venom factor: structure, function, and humanization for therapeutic complement depletion. *Toxicon*, 56(7), 1198-1222.

[71] Zeng, L., Sun, Q. Y., Jin, Y., Zhang, Y., Lee, W. H., & Zhang, Y. (2012). Molecular cloning and characterization of a complement-depleting factor from king cobra, Ophiophagus hannah. *Toxicon*, 60(3), 290-301.

[72] Goodman, S. L., & Picard, M. (2012). Integrins as therapeutic targets. *Trends in Pharmacological Sciences*, 33(7), 405-412.

[73] Yeh, C. H., Peng, H. C., & Huang, T. F. (1998). Accutin, a new disintegrin, inhibits angiogenesis in vitro and in vivo by acting as integrin $\alpha(v)\beta3$ antagonist and inducing apoptosis. *Blood*, 92(9), 3268-3276.

[74] Kang, I. C., Kim, DS, Jang, Y., & Chung, K. H. (2000). Suppressive mechanism of salmosin, a novel disintegrin in B16 melanoma cell metastasis. *Biochemical and Biophysical Research Communications*, 275(1), 169-173.

[75] Minea, R., Swenson, S., Costa, F., Chen, T. C., & Markland, F. S. (2006). Development of a novel recombinant disintegrin, contortrostatin, as an effective anti-tumor and anti-angiogenic agent. *Pathophysiology of Haemostasis and Thrombosis*, 34(4-5), 177-183.

[76] Zhou, X. D., Jin, Y., Chen, R. Q., Lu, Q. M., Wu, J. B., Wang, W. Y., & Xiong, Y. L. (2004). Purification, cloning and biological characterization of a novel disintegrin from Trime resurus jerdonii venom. *Toxicon* , 43(1), 69-75.

[77] Galán, J. A., Sánchez, E. E., Rodríguez-Acosta, A., Soto, J. G., Bashir, S., Mc Lane, M. A., Paquette-Straub, C., & Pérez, J. C. (2008). Inhibition of lung tumor colonization and cell migration with the disintegrin crotatroxin 2 isolated from the venom of Crotalus atrox. *Toxicon*, 51(7), 1186-1196.

[78] Yeh, C. H., Peng, H. C., Yang, R. S., & Huang, T. F. (2001). Rhodostomin, a snake venom disintegrin, inhibits angiogenesis elicited by basic fibroblast growth factor and suppresses tumor growth by a selective $\alpha v/\beta3$ blockade of endothelial cells. *Molecular Pharmacology*, 59(5), 133-1342.

[79] Thangam, R., Gunasekaran, P., Kaveri, K., Sridevi, G., Sundarraj, S., Paulpandi, M., & Kannan, S. (2012). A novel disintegrin protein from Naja naja venom induces cytotoxicity and apoptosis in human cancer cell lines in vitro. *Process Biochemistry*, 47(8), 1243-1249.

[80] Guo, C., Liu, S., Yao, Y., Zhang, Q., & Sun, M. Z. (2012). Past decade study of snake venom L-amino acid oxidase. *Toxicon*, 60(3), 302-311.

[81] Ahn, M. Y., Lee, B. M., & Kim, Y. S. (1997). Characterization and cytotoxicity of L-amino acid oxidase from the venom of king cobra (Ophiophagus hannah). *International Journal of Biochemistry and Cell Biology*, 29(6), 911-919.

[82] Zhang, L., & Wu, W. T. (2008). Isolation and characterization of ACTX-6: a cytotoxic L-amino acid oxidase from Agkistrodon acutus snake venom. *Natural Product Research*, 22(6), 554-563.

[83] Sun, L. K., Yoshii, Y., Hyodo, A., Tsurushima, H., Saito, A., Harakuni, T., Li, Y. P., Kariya, K., Nozaki, M., & Morine, N. (2003). Apoptotic effect in the glioma cells induced by specific protein extracted from Okinawa Habu (Trimeresurus flavoviridis) venom in relation to oxidative stress. *Toxicology in Vitro*, 17(2), 169-177.

[84] Naumann, G. B., Silva, L. F., Silva, L., Faria, G., Richardson, M., Evangelista, K., Kohlhoff, M., Gontijo, C. M. F., Navdaev, A., Rezende, F. F., Eble, J. A., & Sanchez, E.

F. (2011). Cytotoxicity and inhibition of platelet aggregation caused by an L-amino acid oxidase from Bothrops leucurus venom. *Biochimica et Biophysica Acta (BBA)- General Subjects*, 1810(7), 683-694.

[85] Daniele, J. J., Bianco, I. D., Delgado, C., Carrillo, D. B., & Fidelio, G. D. (1997). A new phospholipase A_2 isoform isolated from Bothrops neuwiedii (Yarara chica) venom with novel kinetic and chromatographic properties. *Toxicon*, 35(8), 1205-1215.

[86] Costa, T. R., Menaldo, D. L., Oliveira, C. Z., Santos-Filho, N. A., Teixeira, S. S., Nomizo, A., Fuly, A. L., Monteiro, M. C., De Souza, B. M., Palma, M. S., Stábeli, R. G., Sampaio, S. V., & Soares, A. M. (2008). Myotoxic phospholipases A2 isolated from Bothrops brazili snake venom and synthetic peptides derived from their C-terminal region: cytotoxic effect on microorganism and tumor cells. *Peptides*, 29(10), 1645-1656.

[87] Rudrammaji, L. M. S., & Gowda, T. V. (1998). Purification and characterization of three acidic, cytotoxic phospholipases A2 from Indian cobra (Naja naja naja) venom. *Toxicon*, 36(6), 921-932.

[88] Soares, M. A., Pujatti, P. B., Fortes-Dias, C. L., Antonelli, L., & Santos, R. G. (2010). Crotalus durissus terrificus venom as a source of antitumoral agents. *Journal of Venomous Animals and Toxins Including Tropical Diseases*, 16(3), 480-492.

[89] Nolte, S., Damasio, D. C., Baréa, A. C., Gomes, J., Magalhães, A., Zischler, L. F. C. M., Stuelp-Campelo, P. M., Elífio-Esposito, S. L., Roque-Barreira, M. C., Reis-Amaral, CA, & Moreno, A. N. (2012). BJcuL, a lectin purified from Bothrops jararacussu venom, induces apoptosis in human gastric carcinoma cells accompanied by inhibition of cell adhesion and actin cytoskeleton disassembly. *Toxicon*, 59(1), 81-85.

[90] Gabriel, L. M., Sanchez, E. F., Silva, S. G., & dos Santos, R. G. (2012). Tumor cytotoxicity of leucurolysin-B, a P-III snake venom metalloproteinase from Bothrops leucurus. *Journal of Venomous Animals and Toxins Including Tropical Diseases*, 18(1), 24-33.

[91] Nunes, E. S., Souza, M. A. A., Vaz, A. F. M., Silva, T. G., Aguiar, J. S., Batista, A. M., Guerra, M. M. P., Guarnieri, M. C., Coelho, L. C. B. B., & Correia, M. T. S. (2012). Cytotoxic effect and apoptosis induction by Bothrops leucurus venom lectin on tumor cell lines. *Toxicon*, 59(7-8), 667-671.

[92] Dehesa-Dávila, M., Martin, BM, Nobile, M., Prestipino, G., & Possani, L. D. (1994). Isolation of a toxin from Centruroides infamatus infamatus Koch scorpion venom that modifies Na^+ permeability on chick dorsal root ganglion cells. *Toxicon*, 32(12), 1487-1493.

[93] Chowell, G., Díaz-Dueñas, P., Bustos-Saldaña, R., Mireles, AA , & Fet, V. (2006). Epidemiological and clinical characteristics of scorpionism in Colima, Mexico (2000-2001). *Toxicon*, 47(7), 753-758.

[94] Goudet, C., Chi, C. W., & Tytgat, J. (2002). An overview of toxins and genes from the venom of the Asian scorpion Buthus martensi Karsch. *Toxicon*, 40(9), 1239-1258.

[95] Ferreira, L. A. F., Alves, E. W., & Henriques, O. B. (1993). Peptide T, a novel bradykinin potentiator isolated from Tityus serrulatus scorpion venom. *Toxicon*, 31(8), 941-947.

[96] Pimenta, A. M. C., & De Lima, M. E. (2005). Small peptides, big world: Biotechnological potential in neglected bioactive peptides from arthropod venoms. *Journal of Peptide Science*, 11(11), 670-676.

[97] Zeng, X. C., Li, W. X., Peng, F., & Zhu, Z. H. (2000). Cloning and characterization of a novel cDNA sequence encoding the precursor of a novel venom peptide (BmKbpp) related to a bradykinin- potentiating peptide from Chinese scorpion Buthus martensii Karsch. *IUBMB Life*, 49(3), 207-210.

[98] El -Saadani, M. A. M., & El -Sayed, M. F. (2003). A bradykinin potentiating peptide from Egyptian cobra venom strongly affects rat atrium contractile force and cellular calcium regulation. *Comparative Biochemistry and Physiology Part C: Toxicology and Pharmacology*, 136(4), 387-395.

[99] Bulet, P., Stöcklin, R., & Menin, L. (2004). Anti-microbial peptides: From invertebrates to vertebrates. *Immunological Reviews*, 198-169.

[100] Elgar, D., Du Plessis, J., & Du Plessis, L. (2006). Cysteine-free peptides in scorpion venom: Geographical distribution, structure-function relationship and mode of action. *African Journal of Biotechnology*, 5(25), 2495-2502.

[101] Corzo, G., Escoubas, P., Villegas, E., Barnham, K. J., He, W., Norton, R. S., & Nakajima, T. (2001). Characterization of unique amphipathic antimicrobial peptides from venom of the scorpion Pandinus imperator. *Biochemical Journal*, 359(1), 35-45.

[102] Cociancich, S., Goyffon, M., Bontems, F., Bulet, P., Bouet, F., Menez, A., & Hoffmann, J. (1993). Purification and Characterization of a Scorpion Defensin, a 4kDa Antibacterial peptide presenting structural similarities with insect defensins and scorpion toxins. *Biochemical and Biophysical Research Communications*, 194(1), 17-22.

[103] Ehret-Sabatier, L., Loew, D., Goyffon, M., Fehlbaum, P., Hoffmann, J. A., Van Dorsselaer, A., & Bulet, P. (1996). Characterization of novel cysteine-rich antimicrobial peptides from scorpion blood. *Journal of Biological Chemistry*, 271(47), 29537-29544.

[104] Conde, R., Zamudio, F. Z., Rodríguez, M. H., & Possani, L. D. (2000). Scorpine, an anti-malaria and anti-bacterial agent purified from scorpion venom. *FEBS Letters*, 471(2-3), 165-168.

[105] Dai, L., Corzo, G., Naoki, H., Andriantsiferana, M., & Nakajima, T. (2002). Purification, structure-function analysis, and molecular characterization of novel linear peptides from scorpion Opisthacanthus madagascariensis. *Biochemical and Biophysical Research Communications*, 293(5), 1514-1522.

[106] Moerman, L., Bosteels, S., Noppe, W., Willems, J., Clynen, E., Schoofs, L., Thevissen, K., Tytgat, J., Van Eldere, J., Van Der Walt, J., & Verdonck, F. (2002). Antibacterial

and antifungal properties of α-helical, cationic peptides in the venom of scorpions from southern Africa. *European Journal of Biochemistry*, 269(19), 4799-4810.

[107] Rodríguez La, Vega. R. C., García, B. I., D'Ambrosio, C., Diego-García, E., Scaloni, A., & Possani, L. D. (2004). Antimicrobial peptide induction in the haemolymph of the Mexican scorpion Centruroides limpidus limpidus in response to septic injury. *Cellular and Molecular Life Sciences*, 61(12), 1507-1519.

[108] Uawonggul, N., Thammasirirak, S., Chaveerach, A., Arkaravichien, T., Bunyatratchata, W., Ruangjirachuporn, W., Jearranaiprepame, P., Nakamura, T., Matsuda, M., Kobayashi, M., Hattori, S., & Daduang, S. (2007). Purification and characterization of Heteroscorpine-1 (HS-1) toxin from Heterometrus laoticus scorpion venom. *Toxicon*, 49(1), 19-29.

[109] Elgar, D., Verdonck, F., Grobler, A., Fourie, C., & Du, Plessis. J. (2006). Ion selectivity of scorpion toxin-induced pores in cardiac myocytes. *Peptides*, 27(1), 55-61.

[110] Torres-Larios, A., Gurrola, G. B., Zamudio, F. Z., & Possani, L. D. (2000). Hadrurin, a new antimicrobial peptide from the venom of the scorpion Hadrurus aztecus. *European Journal of Biochemistry*, 267(16), 5023-5031.

[111] Díaz, P., D'Suze, G., Salazar, V., Sevcik, C., Shannon, Sherman. N. E., & Fox, J. W. (2009). Antibacterial activity of six novel peptides from Tityus discrepans scorpion venom. A fluorescent probe study of microbial membrane Na$^+$ permeability changes. *Toxicon*, 54(6), 802-817.

[112] Shijin, Y., Hong, Y., Yibao, M., Zongyun, C., Han, S., Yingliang, W., Zhijian, C., & Wenxin, L. (2008). Characterization of a new Kv1.3 channel-specific blocker, J123, from the scorpion Buthus martensii Karsch. *Peptides*, 29(9), 1514-1520.

[113] Wang, Z., Wang, W., Shao, Z., Gao, B., Li, J., Che, J. H., & Zhang, W. (2009). Eukaryotic expression and purification of anti-epilepsy peptide of Buthus martensii Karsch and its protein interactions. *Molecular and Cellular Biochemistry*, 330(1-2), 97-104.

[114] Shao, J., Kang, N., Liu, Y., Song, S., Wu, C., & Zhang, J. (2007). Purification and characterization of an analgesic peptide from Buthus martensii Karsch. *Biomedical Chromatography*, 21(12), 1266-1271.

[115] Bai, Z. T., Liu, T., Pang, X. Y., Chai, Z. F., & Ji, Y. H. (2007). Suppression by intrathecal BmK IT2 on rat spontaneous pain behaviors and spinal c-Fos expression induced by formalin. *Brain Research Bulletin*, 73(4-6), 248-253.

[116] Gupta, S. D., Gomes, A., Debnath, A., & Saha, A. (2010). Apoptosis induction in human leukemic cells by a novel protein Bengalin, isolated from Indian black scorpion venom: Through mitochondrial pathway and inhibition of heat shock proteins. *Chemico-Biological Interactions*, 183(2), 293-303.

[117] Deshane, J., Garner, C. C., & Sontheimer, H. (2003). Chlorotoxin inhibits glioma cell invasion via matrix metalloproteinase-2. *Journal of Biological Chemistry*, 278(6), 4135-4144.

[118] Pessini, A. C., Kanashiro, A., Malvar, Dd. C., Machado, R. R., Soares, D. M., Figueiredo, MJ, Kalapothakis, E., & Souza, G. E. P. (2008). Inflammatory mediators involved in the nociceptive and oedematogenic responses induced by Tityus serrulatus scorpion venom injected into rat paws. *Toxicon*, 52(7), 729-736.

[119] Jaravine, V. A., Nolde, D. E., Reibarkh, M. J., Korolkova, Y. V., Kozlov, S. A., Pluzhnikov, K. A., Grishin, E. V., & Arseniev, A. S. (1997). Three-dimensional structure of toxin OSK1 from Orthochirus scrobiculosus scorpion venom. *Biochemistry*, 36(6), 1223-1232.

[120] Garcia-Calvo, M., Leonard, R. J., Novick, J., Stevens, S. P., Schmalhofer, W., Kaczorowski, G. J., & Garcia, M. L. (1993). Purification, characterization, and biosynthesis of margatoxin, a component of Centruroides margaritatus venom that selectively inhibits voltage-dependent potassium channels. *Journal of Biological Chemistry*, 268(25), 18866-18874.

[121] Crest, M., Jacquet, G., Gola, M., Zerrouk, H., Benslimane, A., Rochat, H., Mansuelle, P., & Martin-Eauclaire, M. F. (1992). Kaliotoxin, a novel peptidyl inhibitor of neuronal BK-type Ca^{2+}-activated K^+ channels characterized from Androctonus mauretanicus mauretanicus venom. *Journal of Biological Chemistry*, 267(3), 1640-1647.

[122] Garcia, ML. (1994). Purification and characterization of three inhibitors of voltage-dependent K^+ channels from Leiurus quinquestriatus var. Hebraeus venom. *Biochemistry*, 33(22), 6834-6839.

[123] Koschak, A., Bugianesi, R. M., Mitterdorfer, J., Kaczorowski, G. J., Garcia, M. L., & Knaus, H. G. (1998). Subunit composition of brain voltage-gated potassium channels determined by hongotoxin-1, a novel peptide derived from Centruroides limbatus venom. *Journal of Biological Chemistry*, 273(5), 2639-2644.

[124] Domingos, Possani. L., Martin, BM, & Svendsen, I. B. (1982). The primary structure of noxiustoxin: A K^+ channel blocking peptide, purified from the venom of the scorpion Centruroides noxius Hoffmann. *Carlsberg Research Communications*, 47(5), 285-289.

[125] Olamendi-Portugal, T., Gómez-Lagunas, F., Gurrola, G. B., & Possani, L. D. (1996). A novel structural class of K^+-channel blocking toxin from the scorpion Pandinus imperator. *Biochemical Journal*, 315(3), 977-981.

[126] Soroceanu, L., Manning Jr, T. J., & Sontheimer, H. (1999). Modulation of glioma cell migration and invasion using Cl^- and K^+ ion channel blockers. *Journal of Neuroscience*, 19(14), 5942-5954.

[127] Mamelak, A. N., & Jacoby, D. B. (2007). Targeted delivery of antitumoral therapy to glioma and other malignancies with synthetic chlorotoxin (TM-601). *Expert Opinion on Drug Delivery*, 4(2), 175-186.

[128] Jacoby, D. B., Dyskin, E., Yalcin, M., Kesavan, K., Dahlberg, W., Ratliff, J., Johnson, E. W., & Mousa, S. A. (2010). Potent pleiotropic anti-angiogenic effects of TM601, a synthetic chlorotoxin peptide. *Anticancer Research*, 30(1), 39-46.

[129] Wu, J. J., Dai, L., Lan, Z. D., & Chi, C. W. (2000). The gene cloning and sequencing of Bm-12, a Chlorotoxin-like peptide from the scorpion Buthus martensi Karsch. *Toxicon*, 38(5), 661-668.

[130] Fu, Y. J., Yin, L. T., Liang, A. H., Zhang, C. F., Wang, W., Chai, B. F., Yang, J. Y., & Fan, X. J. (2007). Therapeutic potential of chlorotoxin-like neurotoxin from the Chinese scorpion for human gliomas. *Neuroscience Letters*, 412(1), 62-67.

[131] D'Suze, G., Rosales, A., Salazar, V., & Sevcik, C. (2010). Apoptogenic peptides from Tityus discrepans scorpion venom acting against the SKBR3 breast cancer cell line. *Toxicon*, 56(8), 1497-1505.

[132] De Lima, ME, Figueiredo, S. G., Pimenta, A. M. C., Santos, D. M., Borges, M. H., Cordeiro, M. N., Richardson, M., Oliveira, L. C., Stankiewicz, M., & Pelhate, M. (2007). Peptides of arachnid venoms with insecticidal activity targeting sodium channels. *Comparative Biochemistry and Physiology Part C: Toxicology and Pharmacology*, 146(1-2), 264-279.

[133] Eitan, M., Fowler, E., Herrmann, R., Duval, A., Pelhate, M., & Zlotkin, E. (1990). A scorpion venom neurotoxin paralytic to insects that affects sodium current inactivation: purification, primary structure, and mode of action. *Biochemistry*, 29(25), 5941-5947.

[134] Karbat, I., Frolow, F., Froy, O., Gilles, N., Cohen, L., Turkov, M., Gordon, D., & Gurevitz, M. (2004). Molecular basis of the high insecticidal potency of scorpion alpha-toxins. *The Journal of Biological Chemistry*, 279(30), 31679-31686.

[135] Borchani, L., Stankiewicz, M., Kopeyan, C., Mansuelle, P., Kharrat, R., Cestèle, S., Karoui, H., Rochat, H., Pelhate, M., & El Ayeb, M. (1997). Purification, structure and activity of three insect toxins from Buthus occitanus tunetanus venom. *Toxicon*, 35(3), 365-382.

[136] Arnon, T., Potikha, T., Sher, D., Elazar, M., Mao, W., Tal, T., Bosmans, F., Tytgat, J., Ben-Arie, N., & Zlotkin, E. (2005). BjαIT: a novel scorpion α-toxin selective for insects- unique pharmacological tool. *Insect Biochemistry and Molecular Biology*, 35(3), 187-195.

[137] Jalali, A., Bosmans, F., Amininasab, M., Clynen, E., Cuypers, E., Zaremirakabadi, A., Sarbolouki, M. N., Schoofs, L., Vatanpour, H., & Tytgat, J. (2005). OD1, the first toxin isolated from the venom of the scorpion Odonthobuthus doriae active on voltage-gated Na+ channels. *FEBS Letters*, 579(19), 4181-4186.

[138] Krimm, I., Gilles, N., Sautière, P., Stankiewicz, M., Pelhate, M., Gordon, D., & Lancelin-M, J. (1999). NMR structures and activity of a novel α-like toxin from the scorpion Leiurus quinquestriatus hebraeus. *Journal of Molecular Biology*, 285(4), 1749-1763.

[139] Cestèle, S., Stankiewicz, M., Mansuelle, P., De Waard, M., Dargent, B., Gilles, N., Pelhate, M., Rochat, H., Martin-Eauclaire-F, M., & Gordon, D. (1999). Scorpion α-like toxins, toxic to both mammals and insects, differentially interact with receptor site 3

on voltage-gated sodium channels in mammals and insects. *European Journal of Neuroscience*, 11(3), 975-985.

[140] Wang, C. G., Ling, M. H., Chi, C. W., Wang, D. C., Stankiewicz, M., & Pelhate, M. (2003). Purification of two depressant insect neurotoxins and their gene cloning from the scorpion Buthus martensi Karsch. The Journal of Peptide Research , 61(1), 7-16.

[141] Hamon, A., Gilles, N., Sautière, P., Martinage, A., Kopeyan, C., Ulens, C., Tytgat, J., Lancelin, J. M., & Gordon, D. (2002). Characterization of scorpion α-like toxin group using two new toxins from the scorpion Leiurus quinquestriatus hebraeus. *European Journal of Biochemistry*, 269(16), 3920-3933.

[142] Pelhate, M., & Zlotkin, E. (1982). Actions of insect toxin and other toxins derived from the venom of the scorpion Androctonus australis on isolated giant axons of the cockroach (Periplaneta americana). *The Journal of Experimental Biology*, 97-67.

[143] Pelhate, M., Stankiewicz, M., & Ben, Khalifa. R. (1998). Anti-insect scorpion toxins: historical account, activities and prospects. *Comptes Rendus des Séances de la Société de Biologie et de Ses Filiales*, 192(3), 463-484.

[144] Escoubas, P., Stankiewicz, M., Takaoka, T., Pelhate, M., Romi-Lebrun, R., Wu, F. Q., & Nakajima, T. (2000). Sequence and electrophysiological characterization of two insect-selective excitatory toxins from the venom of the Chinese scorpion Buthus martensi. *FEBS Letters*, 483(2-3), 175-80.

[145] Ori, M., & Ikeda, H. (1998). Spider venoms and spider toxins. *Journal of Toxicology-Toxin Reviews*, 17(3), 405-426.

[146] Rash, L. D., & Hodgson, W. C. (2001). Pharmacology and biochemistry of spider venoms. *Toxicon*, 40(3), 225-254.

[147] Bode, F., Sachs, F., & Franz, M. R. (2001). Tarantula peptide inhibits atrial fibrillation. *Nature*, 409(6816), 35-36.

[148] Mc Glasson, D. L., Babcock, J. L., Berg, L., & Triplett, D. A. (1993). ARACHnase: An evaluation of a positive control for platelet neutralization procedure testing with seven commercial activated partial thromboplastin time reagents. *American Journal of Clinical Pathology*, 100(5), 576-578.

[149] Futrell, J. M. (1992). Loxoscelism. *American Journal of the Medical Sciences*, 304(4), 261-267.

[150] Yan, L., & Adams, M. E. (1998). Lycotoxins, antimicrobial peptides from venom of the wolf spider Lycosa carolinensis. *Journal of Biological Chemistry*, 273(4), 2059-2066.

[151] Mazzuca, M., Heurteaux, C., Alloui, A., Diochot, S., Baron, A., Voilley, N., Blondeau, N., Escoubas, P., Gelot, A., Cupo, A., Zimmer, A., Zimmer, A. M., Eschalier, A., & Lazdunski, M. (2007). A tarantula peptide against pain via ASIC1a channels and opioid mechanisms. *Nature Neuroscience*, 10(8), 943-945.

[152] Van Meeteren, L. A., Frederiks, F., Giepmans, B. N. G., Fernandes, Pedrosa. M. F., Billington, S. J., Jost, B. H., Tambourgi, D. V., & Moolenaar, W. H. (2004). Spider and bacterial Sphingomyelinases D target cellular lysophosphatidic acid receptors by hydrolyzing lysophosphatidylcholine. *Journal of Biological Chemistry*, 279(12), 10833-10836.

[153] Fernandes, Pedrosa. M. F., Junqueira de, Azevedo. I. D. L. M., Gonçalves-de-Andrade, R. M., Van Den, Berg. C. W., Ramos, C. R. R., Lee, Ho. P., & Tambourgi, D. V. (2002). Molecular cloning and expression of a functional dermonecrotic and haemolytic factor from Loxosceles laeta venom. *Biochemical and Biophysical Research Communications*, 298(5), 638-645.

[154] Murakami, M. T., Fernandes-Pedrosa, M. F., Tambourgi, D. V., & Arni, R. K. (2005). Structural basis for metal ion coordination and the catalytic mechanism of sphingomyelinases D. *Journal of Biological Chemistry*, 280(14), 13658-13664.

[155] De Almeida, D. M., Fernandes-Pedrosa, M. F., Gonçalves de, Andrade. R. M., Marcelino, J. R., Gondo-Higashi, H., Junqueira de, Azevedo. I. D. L. M., Ho, P. L., Van Den, Berg. C., & Tambourgi, D. V. (2008). A new anti-loxoscelic serum produced against recombinant sphingomyelinase D: Results of preclinical trials. *American Journal of Tropical Medicine and Hygiene*, 79(3), 463-470.

[156] Bubien, J. K., Ji, H. L., Gillespie, G. Y., Fuller, C. M., Markert, J. M., Mapstone, T. B., & Benos, D. J. (2004). Cation selectivity and inhibition of malignant glioma Na+ channels by Psalmotoxin 1. American Journal of Physiology- Cell Physiology 56-5) , 287(5), C1282-C1291.

[157] Rodrigues, E. G., Dobroff, A. S. S., Cavarsan, C. F., Paschoalin, T., Nimrichter, L., Mortara, R. A., Santos, E. L., Fázio, MA, Miranda, A., Daffre, S., & Travassos, L. R. (2008). Effective topical treatment of subcutaneous murine B16F10-Nex2 melanoma by the antimicrobial peptide gomesin. *Neoplasia*, 10(1), 61-68.

[158] Khan, S. A., Zafar, Y., Briddon, R. W., Malik, K. A., & Mukhtar, Z. (2006). Spider venom toxin protects plants from insect attack. *Transgenic Research*, 15(3), 349-357.

[159] De Castro, C. S., Silvestre, F. G., Araújo, S. C., Yazbeck, G. D. M., Mangili, O. C., Cruz, I., Chávez-Olórtegui, C., & Kalapothakis, E. (2004). Identification and molecular cloning of insecticidal toxins from the venom of the brown spider Loxosceles intermedia. *Toxicon*, 44(3), 273-280.

[160] Figueiredo, S. G., Garcia, M. E. L. P., Valentim, A. D. C., Cordeiro, M. N., Diniz, C. R., & Richardson, M. (1995). Purification and amino acid sequence of the insecticidal neurotoxin Tx4(6-1) from the venom of the'armed' spider Phoneutria nigriventer (Keys). *Toxicon*, 33(1), 83-93.

[161] Zobel-Thropp, P. A., Kerins, A. E., & Binford, G. J. (2012). Sphingomyelinase D in sicariid spider venom is a potent insecticidal toxin. *Toxicon*, 60(3), 265-271.

[162] Ferrat, G., Bosmans, F., Tytgat, J., Pimentel, C., Chagot, B., Gilles, N., Nakajima, T., Darbon, H., & Corzo, G. (2005). Solution structure of two insect-specific spider toxins and their pharmacological interaction with the insect voltage-gated Na$^+$ channel. *Proteins Structure, Function and Genetics*, 59(2), 368-379.

[163] Toledo, R. C., & Jared, C. (1995). Cutaneous granular glands and amphibian venoms. *Comparative Biochemistry and Physiology Part A: Physiology*, 111(1), 1-29.

[164] Tang, Y. Q., Tian, Sh., Hua, Jc., Wu, Sx., Zou, G., Wu, Gf., & Zhao, Em. (1990). Isolation, chemical and biological characterization of margaratensin, a neurotensin-related peptide from the skin of Rana margaratae. *Science in China (Scientia Sinica) Series B*, 33(7), 828-834.

[165] Wang, D. L., Qi, F. H., Tang, W., & Wang, F. S. (2011). Chemical constituents and bioactivities of the skin of Bufo bufo gargarizans cantor. *Chemistry and Biodiversity*, 8(4), 559-567.

[166] Cruz, J. S., & Matsuda, H. (1993). Arenobufagin, a compound in toad venom, blocks Na$^+$-K$^+$ pump current in cardiac myocytes. *European Journal of Pharmacology*.

[167] Shindelman, J., Mosher, H. S., & Fuhrman, F. A. (1969). Atelopidtoxin from the Panamanian frog, Atelopus zeteki. *Toxicon*, 7(4), 315-319.

[168] Erspamer, G. F., Severini, C., Erspamer, V., & Melchiorri, P. (1989). Pumiliotoxin B-like alkaloid in extracts of the skin of the australian myobatrachid frog Pseudophryne coriacea: effects on the systemic blood pressure of experimental animals and the rat heart. *Neuropharmacology*, 28(4), 319-328.

[169] Zhang, Y., Yu, G., Wang, Y., Zhang, J., Wei, S., Lee, W., & Zhang, Y. (2010). A novel annexin A2 protein with platelet aggregation-inhibiting activity from amphibian Bombina maxima skin. *Toxicon*, 56(3), 458-465.

[170] Cunha, Filho. G. A., Schwartz, C. A., Resck, I. S., Murta, M. M., Lemos, S. S., Castro, M. S., Kyaw, C., Pires Jr, O. R., Leite, J. R. S., Bloch Jr, C., & Schwartz, E. F. (2005). Antimicrobial activity of the bufadienolides marinobufagin and telocinobufagin isolated as major components from skin secretion of the toad Bufo rubescens. *Toxicon*, 45(6), 777-782.

[171] Dourado, F. S., Leite, J. R. S. A., Silva, L. P., Melo, J. A. T., Bloch Jr, C., & Schwartz, E. F. (2007). Antimicrobial peptide from the skin secretion of the frog Leptodactylus syphax. *Toxicon*, 50(4), 572-580.

[172] Preusser, H. J., Habermehl, G., Sablofski, M., & Schmall, Haury. D. (1975). Antimicrobial activity of alkaloids from amphibian venoms and effects on the ultrastructure of yeast cells. *Toxicon*, 13(4), 285-289.

[173] Cui, X., Inagaki, Y., Xu, H., Wang, D., Qi, F., Kokudo, N., Fang, D., & Tang, W. (2010). Anti-hepatitis B virus activities of cinobufacini and its active components bufalin and cinobufagin in HepG2.2.15 Cells. *Biological and Pharmaceutical Bulletin*, 33(10), 1728-1732.

[174] Tempone, A. G., Pimenta, D. C., Lebrun, I., Sartorelli, P., Taniwaki, N. N., de Andrade Jr, H. F., Antoniazzi, MM, & Jared, C. (2008). Antileishmanial and antitrypanosomal activity of bufadienolides isolated from the toad Rhinella jimi parotoid macrogland secretion. *Toxicon*, 52(1), 13-21.

[175] Yogeeswari, P., Sriram, D., Bal, T. R., & Thirumurugan, R. (2006). Epibatidine and its analogues as nicotinic acetylcholine receptor agonist: an update. *Natural Product Research*, 20(5), 497-505.

[176] Erspamer, V., Melchiorri, P., Falconieri-Erspamer, G., Negri, L., Corsi, R., Severini, C., Barra, D., Simmaco, M., & Kreil, G. (1989). Deltrophins: a family of naturally occurring peptides with high affinity and selectivity for δ opioid binding sites. *Proceedings of the National Academy of Sciences of the United States of America*, 86(13), 5188-5192.

[177] Wang, X. L., Zhao, G. H., Zhang, J., Shi, Q. Y., Guo, W. X., Tian, X. L., Qiu, J. Z., Yin, L. Z., Deng, X. M., & Song, Y. (2011). Immunomodulatory effects of cinobufagin isolated from Chan Su on activation and cytokines secretion of immunocyte in vitro. *Journal of Asian Natural Products Research*, 13(5), 383-392.

[178] Qi, F., Inagaki, Y., Gao, B., Cui, X., Xu, H., Kokudo, N., Li, A., & Tang, W. (2011). Bufalin and cinobufagin induce apoptosis of human hepatocellular carcinoma cells via Fas- and mitochondria-mediated pathways. *Cancer Science*, 102(5), 951-958.

[179] Su, C. L., Lin, T. Y., Lin, C. N., & Won, S. J. (2009). Involvement of caspases and apoptosis-inducing factor in bufotalin-induced apoptosis of hep 3B cells. *Journal of Agricultural and Food Chemistry*, 57(1), 55-61.

[180] Balboni, F., Bernabei, P. A., Barberio, C., Sanna, A., Rossi, Ferrini. P., & Delfino, G. (1992). Cutaneous venom of Bombina variegata pachypus (Amphibia, anura): effects on the growth of the human HL 60 cell line. *Cell Biology International Reports*, 16(4), 329-338.

[181] Ghavami, S., Asoodeh, A., Klonisch, T., Halayko, A. J., Kadkhoda, K., Kroczak, T. J., Gibson, S. B., Booy, E. P., Naderi-Manesh, H., & Los, M. (2008). Brevinin-2R1 semi-selectively kills cancer cells by a distinct mechanism, which involves the lysosomal-mitochondrial death pathway. *Journal of Cellular and Molecular Medicine*, 12(3), 1005-1022.

[182] Marenah, L., Shaw, C., Orr, D. F., Mc Clean, S., Flatt, P. R., & Abdel-Wahab, Y. H. A. (2004). Isolation and characterisation of an unexpected class of insulinotropic peptides in the skin of the frog Agalychnis litodryas. *Regulatory Peptides*, 120(1-3), 33-38.

[183] Marsh, N. A., & Whaler, B. C. (1980). The effects of honey bee (Apis mellifera L.) venom and two of its constituents, melittin and phospholipase A2, on the cardiovascular system of the rat. *Toxicon*, 18(4), 427-435.

[184] Kaplinsky, E., Ishay, J., Ben-Shachar, D., & Gitter, S. (1977). Effects of bee (Apis mellifera) venom on the electrocardiogram and blood pressure. *Toxicon*, 15(3), 251-256.

[185] Vick, J. A., Shipman, W. H., & Brooks, R. Jr. (1974). Beta adrenergic and anti-arrhythmic effects of cardiopep, a newly isolated substance from whole bee venom. *Toxicon*, 12(2), 139-144.

[186] Ho, C. L., Hwang, L. L., Lin, Y. L., Chen, C. T., Yu, H. M., & Wang, K. T. (1994). Cardiovascular effects of mastoparan B and its structural requirements. *European Journal of Pharmacology*, 259(3), 259-264.

[187] Choo, Y. M., Lee, K. S., Yoon, H. J., Qiu, Y., Wan, H., Sohn, M. R., Sohn, H. D., & Jin, B. R. (2012). Antifibrinolytic role of a bee venom serine protease inhibitor that acts as a plasmin Inhibitor. *PLoS ONE*, 7(2).

[188] Choo, Y. M., Lee, K. S., Yoon, H. J., Kim, B. Y., Sohn, M. R., Roh, J. Y., Je, Y. H., Kim, N. J., Kim, I., Woo, S. D., Sohn, H. D., & Jin, B. R. (2010). Dual function of a bee venom serine protease: prophenoloxidase-activating factor in arthropods and fibrin(ogen)olytic enzyme in mammals. *PLoS ONE*, 5(5), 1.

[189] Qiu, Y., Choo, Y. M., Yoon, H. J., Jia, J., Cui, Z., Wang, D., Kim, D. H., Sohn, H. D., & Jin, B. R. (2011). Fibrin(ogen)olytic activity of bumblebee venom serine protease. *Toxicology and Applied Pharmacology*, 255(2), 207-213.

[190] Qiu, Y., Choo, Y. M., Yoon, H. J., & Jin, B. R. (2012). Molecular cloning and fibrin(ogen)olytic activity of a bumblebee (Bombus hypocrita sapporoensis) venom serine protease. *Journal of Asia-Pacific Entomology*, 15(1), 79-82.

[191] Han, J., You, D., Xu, X., Han, W., Lu, Y., Lai, R., & Meng, Q. (2008). An anticoagulant serine protease from the wasp venom of Vespa magnifica. *Toxicon*, 51(5), 914-922.

[192] Haim, B., Rimon, A., Ishay, J. S., & Rimon, S. (1999). Purification, characterization and anticoagulant activity of a proteolytic enzyme from Vespa orientalis venom. *Toxicon*, 37(5), 825-829.

[193] Yang, H., Xu, X., Zhang, D., Lai, K., & , R. (2008). A phospholipase A1 platelet activator from the wasp venom of Vespa magnifica (Smith). *Toxicon*, 51(2), 289-296.

[194] Xu, X., Li, J., Lu, Q., Yang, H., Zhang, Y., & Lai, R. (2006). Two families of antimicrobial peptides from wasp (Vespa magnifica) venom. *Toxicon*, 47(2), 249-253.

[195] Lazarev, V. N., Parfenova, T. M., Gularyan, S. K., Misyurina, O. Y., Akopian, T. A., & Govorun, V. M. (2002). Induced expression of melittin, an antimicrobial peptide, inhibits infection by Chlamydia trachomatis and Mycoplasma hominis in a HeLa cell line. *International journal of antimicrobial agents*, 19(2), 133-7.

[196] Choo, Y. M., Lee, K. S., Yoon, H. J., Je, Y. H., Lee, S. W., Sohn, H. D., & Jin, B. R. (2010). Molecular cloning and antimicrobial activity of bombolitin, a component of bumblebee Bombus ignitus venom. *Comparative Biochemistry and Physiology Part B: Biochemistry and Molecular Biology*, 156(3), 168-173.

[197] Stöcklin, R., Favreau, P., Thai, R., Pflugfelder, J., Bulet, P., & Mebs, D. (2010). Structural identification by mass spectrometry of a novel antimicrobial peptide from the

venom of the solitary bee Osmia rufa (Hymenoptera: Megachilidae). *Toxicon*, 55(1), 20-27.

[198] Chen, W., Yang, X., Yang, X., Zhai, L., Lu, Z., Liu, J., & Yu, H. (2008). Antimicrobial peptides from the venoms of Vespa bicolor Fabricius. *Peptides*, 29(11), 1887-1892.

[199] Baek, J. H., & Lee, S. H. (2010). Isolation and molecular cloning of venom peptides from Orancistrocerus drewseni (Hymenoptera: Eumenidae). *Toxicon*, 55(4), 711-718.

[200] Yang, E. J., Jiang, J. H., Lee, S. M., Yang, S. C., Hwang, H. S., Lee, M. S., & Choi-M, S. (2010). Bee venom attenuates neuroinflammatory events and extends survival in amyotrophic lateral sclerosis models. *Journal of Neuroinflammation*, 7(1), 69.

[201] Lee, J., Kim, S., Kim, T., Lee, S., Yang, H., Lee, D., & Lee, Y. (2004). Anti-inflammatory effect of bee venom on type II collagen-induced arthritis. *The American journal of Chinese medicine*, 32(3), 361-7.

[202] Park, J. H., Kim, K. H., Lee, W. R., Han, S. M., & Park, K. K. (2012). Protective effect of melittin on inflammation and apoptosis in acute liver failure. *Apoptosis*, 17(1), 61-69.

[203] Shkenderov, S., & Koburova, K. (1982). Adolapin- a newly isolated analgetic and anti-inflammatory polypeptide from bee venom. *Toxicon*, 20(1), 317-321.

[204] Haux, P. (1969). Amino acid sequence of MCD-peptide, a specific mast cell-degranulating peptide from bee venom. *Hoppe-Seyler's Zeitschrift fur Physiologische Chemie*, 350(5), 536-546.

[205] Kettner, A., Henry, H., Hughes, G. J., Corradin, G., & Spertini, F. (1999). IgE and T-cell responses to high-molecular weight allergens from bee venom. *Clinical & Experimental Allergy*, 29(3), 394-401.

[206] Kettner, A., Hughes, G. J., Frutiger, S., Astori, M., Roggero, M., Spertini, F., & Corradin, G. (2001). Api m 6: a new bee venom allergen. *Journal of Allergy and Clinical Immunology*, 107(5), 914-920.

[207] Pantera, B., Hoffman, D. R., Carresi, L., Cappugi, G., Turillazzi, S., Manao, G., Severino, M., Spadolini, I., Orsomando, G., Moneti, G., & Pazzagli, L. (2003). Characterization of the major allergens purified from the venom of the paper wasp Polistes gallicus. *Biochimica et Biophysica Acta (BBA)- General Subjects*, 1623(2-3), 72-81.

[208] An, S., Chen, L., Wei, J. F., Yang, X., Ma, D., Xu, X., He, S., Lu, J., & Lai, R. (2012). Purification and characterization of two new allergens from the venom of Vespa magnifica. *PLoS One*, 7(2), 27.

[209] Trindade, R. A., Kiyohara, P. K., de Araujo, P. S., & Bueno da. Costa, M. H. (2012). PLGA microspheres containing bee venom proteins for preventive immunotherapy. *International Journal of Pharmaceutics*, 423(1), 124-133.

[210] Dani, M. P., & Richards, E. H. (2010). Identification, cloning and expression of a second gene (vpr1) from the venom of the endoparasitic wasp, Pimpla hypochondriaca that displays immunosuppressive activity. *Journal of Insect Physiology*, 56(2), 195-203.

[211] Baptista-Saidemberg, N. B., Saidemberg, D. M., de Souza, B. M., Cesar-Tognoli, L. M., Ferreira, V. M., Mendes, M. A., Cabrera, M. P., Ruggiero, Neto. J., & Palma, M. S. (2010). Protonectin (1-6): a novel chemotactic peptide from the venom of the social wasp Agelaia pallipes pallipes. *Toxicon*, 56(6), 880-889.

[212] Souza, B. M., Mendes, M. A., Santos, L. D., Marques, M. R., Cesar, L. M., Almeida, R. N., Pagnocca, F. C., Konno, K., & Palma, M. S. (2005). Structural and functional characterization of two novel peptide toxins isolated from the venom of the social wasp Polybia paulista. *Peptides*, 26(11), 2157-2164.

[213] Graaf, D. C., Aerts, M., Danneels, E., & Devreese, B. (2009). Bee, wasp and ant venomics pave the way for a component-resolved diagnosis of sting allergy. *Journal of Proteomics*, 72(2), 145-154.

[214] Ownby, C. L., Powell, J. R., Jiang, M. S., & Fletcher, J. E. (1997). Melittin and phospholipase A2 from bee (Apis mellifera) venom cause necrosis of murine skeletal muscle in vivo. *Toxicon*, 35(1), 67-80.

[215] Hait, W. N., Grais, L., Benz, C., & Cadman, E. C. (1985). Inhibition of growth of leukemic cells by inhibitors of calmodulin: phenothiazines and melittin. *Cancer Chemotherapy and Pharmacology*, 14(3), 202-205.

[216] Moon, D. O., Park, S. Y., Heo, M. S., Kim, K. C., Park, C., Ko, W. S., Choi, Y. H., & Kim, G. Y. (2006). Key regulators in bee venom-induced apoptosis are Bcl-2 and caspase-3 in human leukemic U937 cells through downregulation of ERK and Akt. *International Immunopharmacology*, 6(12), 1796-1807.

[217] Orsolic, N., Sver, L., Verstovsek, S., Terzic, S., & Basic, I. (2003). Inhibition of mammary carcinoma cell proliferation in vitro and tumor growth in vivo by bee venom. *Toxicon*, 41(7), 861-870.

[218] Putz, T., Ramoner, R., Gander, H., Rahm, A., Bartsch, G., & Thurnher, M. (2006). Antitumor action and immune activation through cooperation of bee venom secretory phospholipase A_2 and phosphatidylinositol-(3,4)-bisphosphate. *Cancer Immunology Immunotherapy*, 55(11), 1374-1383.

[219] Cerovsky, V., Budesinsky, M., Hovorka, O., Cvacka, J., Voburka, Z., Slaninova, J., Borovickova, L., Fucik, V., Bednarova, L., Votruba, I., & Straka, J. (2009). Lasioglossins: three novel antimicrobial peptides from the venom of the eusocial bee Lasioglossum laticeps (Hymenoptera: Halictidae). *ChemBioChem*, 10(12), 2089-2099.

[220] Heinen, T. E., & da Veiga, A. B. (2011). Arthropod venoms and cancer. *Toxicon*, 57(4), 497-511.

[221] Pfeiffer, D. R., Gudz, T. I., Novgorodov, S. A., & Erdahl, W. L. (1995). The peptide mastoparan is a potent facilitator of the mitochondrial permeability transition. *The Journal of Biological Chemistry*, 270(9), 4923-4932.

[222] Souza, B. M., Silva, A. V., Resende, V. M., Arcuri, H. A., Santos, Cabrera. M. P., Ruggiero, Neto. J., & Palma, M. S. (2009). Characterization of two novel polyfunctional

mastoparan peptides from the venom of the social wasp Polybia paulista. *Peptides*, 30(8), 1387-1395.

[223] Wang, K. R., Zhang, B. Z., Zhang, W., Yan, J. X., Li, J., & Wang, R. (2008). Antitumor effects, cell selectivity and structure-activity relationship of a novel antimicrobial peptide polybia-MPI. *Peptides*, 29(6), 963-968.

[224] Kim, J. Y., Cho, S. H., Kim, Y. W., Jang, E. C., Park, S. Y., Kim, E. J., & Lee, S. K. (1999). Effects of BCG, lymphotoxin and bee venom on insulitis and development of IDDM in non-obese diabetic mice. *J Korean Med Sci*, 14(6), 648-652.

[225] Shin, Y., Moni, R. W., Lueders, J. E., & Daly, J. W. (1994). Effects of the amphiphilic peptides mastoparan and adenoregulin on receptor binding, G proteins, phosphoinositide breakdown, cyclic AMP generation, and calcium influx. *Cell Mol Neurobiol*, 14(2), 133-157.

[226] Straub, S. G., James, R. F., Dunne, M. J., & Sharp, G. W. (1998). Glucose augmentation of mastoparan-stimulated insulin secretion in rat and human pancreatic islets. *Diabetes*, 47(7), 1053-1057.

[227] Mc Gain, F., & Winkel, K. D. (2002). Ant sting mortality in Australia. *Toxicon*, 40(8), 1095-1100.

[228] Malta, M. B., Lira, MS, Soares, S. L., Rocha, G. C., Knysak, I., Martins, R., Guizze, S. P., Santoro, M. L., & Barbaro, K. C. (2008). Toxic activities of Brazilian centipede venoms. *Toxicon*, 52(2), 255-263.

[229] Carrijo-Carvalho, L. C., & Chudzinski-Tavassi, A. M. (2007). The venom of the Lonomia caterpillar: an overview. *Toxicon*, 49(6), 741-757.

[230] Veiga, A. B., Ribeiro, J. M., Guimaraes, J. A., & Francischetti, I. M. (2005). A catalog for the transcripts from the venomous structures of the caterpillar Lonomia obliqua: identification of the proteins potentially involved in the coagulation disorder and hemorrhagic syndrome. *Gene*, 355-11.

[231] Bohrer, C. B., Reck, Junior. J., Fernandes, D., Sordi, R., Guimaraes, J. A., Assreuy, J., & Termignoni, C. (2007). Kallikrein-kinin system activation by Lonomia obliqua caterpillar bristles: involvement in edema and hypotension responses to envenomation. *Toxicon*, 49(5), 663-669.

[232] Donato, J. L., Moreno, R. A., Hyslop, S., Duarte, A., Antunes, E., Le Bonniec, B. F., Rendu, F., & de Nucci, G. (1998). Lonomia obliqua caterpillar spicules trigger human blood coagulation via activation of factor X and prothrombin. *Thromb Haemost*, 79(3), 539-542.

[233] Chudzinski-Tavassi, A. M., Schattner, M., Fritzen, M., Pozner, R. G., Reis, C. V., Lourenco, D., & Lazzari, MA. (2001). Effects of lopap on human endothelial cells and platelets. *Haemostasis*, 31(3-6), 257-265.

[234] Pinto, A. F., Dobrovolski, R., Veiga, A. B., & Guimaraes, J. A. (2004). Lonofibrase, a novel alpha-fibrinogenase from Lonomia obliqua caterpillars. *Thrombosis Research*, 113(2), 147-154.

[235] Alvarez, Flores. M. P., Fritzen, M., Reis, C. V., & Chudzinski-Tavassi, A. M. (2006). Losac, a factor X activator from Lonomia obliqua bristle extract: its role in the pathophysiological mechanisms and cell survival. *Biochemical and Biophysical Research Communications*, 343(4), 1216-1223.

[236] Guerrero, B., Arocha-Pinango, C. L., Salazar, A. M., Gil, A., Sanchez-Acosta, EE, Rodriguez., A., & Lucena, S. (2011). The effects of Lonomin V, a toxin from the caterpillar (Lonomia achelous), on hemostasis parameters as measured by platelet function. *Toxicon*, 58(4), 293-303.

[237] Hink, W. F., Romstedt, K. J., Burke, J. W., Doskotch, R. W., & Feller, D. R. (1989). Inhibition of human platelet aggregation and secretion by ant venom and a compound isolated from venom. *Inflammation*, 13(2), 175-184.

[238] Orivel, J., Redeker, V., Le Caer, J. P., Krier, F., Revol-Junelles, A. M., Longeon, A., Chaffotte, A., Dejean, A., & Rossier, J. (2001). Ponericins, new antibacterial and insecticidal peptides from the venom of the ant Pachycondyla goeldii. *The Journal of Biological Chemistry*, 276(21), 17823-17829.

[239] Zelezetsky, I., Pag, U., Antcheva, N., Sahl, H. G., & Tossi, A. (2005). Identification and optimization of an antimicrobial peptide from the ant venom toxin pilosulin. *Archives of Biochemistry and Biophysics*, 434(2), 358-364.

[240] Peng, K., Kong, Y., Zhai, L., Wu, X., Jia, P., Liu, J., & Yu, H. (2010). Two novel antimicrobial peptides from centipede venoms. *Toxicon*, 55(2-3), 274-279.

[241] Altman, R. D., Schultz, D. R., Collins-Yudiskas, B., Aldrich, J., Arnold, P. I., & Brown, H. E. (1984). The effects of a partially purified fraction of an ant venom in rheumatoid arthritis. *Arthritis & Rheumatism*, 27(3), 277-284.

[242] Pan, J., & Hink, W. F. (2000). Isolation and characterization of myrmexins, six isoforms of venom proteins with anti-inflammatory activity from the tropical ant, Pseudomyrmex triplarinus. *Toxicon*, 38(10), 1403-1413.

[243] Hoffman, D. R. (1993). Allergens in Hymenoptera venom XXIV: the amino acid sequences of imported fire ant venom allergens Sol i II, Sol i III, and Sol i IV. *Journal of Allergy and Clinical Immunology*, 91(1), 71-78.

[244] Hoffman, D. R., Sakell, R. H., & Schmidt, M. (2005). Sol i 1, the phospholipase allergen of imported fire ant venom. *Journal of Allergy and Clinical Immunology*, 115(3), 611-616.

[245] Arbiser, J. L., Kau, T., Konar, M., Narra, K., Ramchandran, R., Summers, S. A., Vlahos, C. J., Ye, K., Perry, B. N., Matter, W., Fischl, A., Cook, J., Silver, P. A., Bain, J., Cohen, P., Whitmire, D., Furness, S., Govindarajan, B., & Bowen, J. P. (2007). Sole-

nopsin, the alkaloidal component of the fire ant (Solenopsis invicta), is a naturally occurring inhibitor of phosphatidylinositol-3-kinase signaling and angiogenesis. *Blood,* 109(2), 560-565.

[246] Sonoda, Y., Hada, N., Kaneda, T., Suzuki, T., Ohshio, T., Takeda, T., & Kasahara, T. (2008). A synthetic glycosphingolipid-induced antiproliferative effect in melanoma cells is associated with suppression of FAK, Akt, and Erk activation. *Biological and Pharmaceutical Bulletin,* 31(6), 1279-1283.

Chapter 2

Peptidomic Analysis of Animal Venoms

Ricardo Bastos Cunha

Additional information is available at the end of the chapter

1. Introduction

The last two decades have witnessed a growing interest in the discovery of new chemical and pharmacological substances of animal origin. Pharmacological tests of toxins obtained from animal venoms revealed its effects on central nervous system, mainly acting on ion channels in heart, intestine, in vascular permeability, etc. Potential applications of these substances have been proposed ranging from human disease treatment to plague control of agricultural interest. In this scenario, the peptidomic analysis has played an increasingly important role.

Venomous organisms are widespread throughout the animal kingdom, comprising more than 100,000 species distributed in all major phyla. Virtually all ecosystems on Earth have venomous or poisonous organisms. Venoms represent an adaptive trait, and an example of convergent evolution. They are truly mortal cocktails, comprising unique mixtures of peptides and proteins naturally tailored by natural selection to operate in defense or attack systems, for the prey or the victim. Venoms represent an enormous reservoir of bioactive compounds able to cure diseases that do not respond to conventional therapies. Darwinian evolution of animal venoms has accumulated in nature a wide variety of biological fluids which resulted in a true combinatorial libraries of hundreds of thousands of molecules potentially active and pharmacologically useful.

Venom is a general term which refers to a variety of toxins used by certain animals that inoculate its victims through a bite, a sting or other sharp body feature. Venoms of vertebrates and invertebrates contain a molecular diversity of proteins and peptides, and other classes of substances, which together form an arsenal of highly effective agents, paralyzing and lethal, mainly used for predation and defense. We must distinguish venom from poison, which is ingested or inhaled by the victim, being absorbed by its digestive system or respiratory system. Animal venoms, in contrast, are administered directly into the lymphatic system, where it acts faster. Only those organisms possessing injection devices (stingers, fangs, spines, hypostomes, spurs or harpoons) which allow the active use of venom for predation can be correctly characterized as venomous. Many other animals secrete lethal substances (insects, centipedes, frogs, fish, etc.), but, as these substances are used primarily for defense purposes, these animals are termed poisonous and cannot be accurately characterized as venomous.

Venomous and poisonous invertebrates include cnidarians [1, 2] (sea anemones, jellyfish and corals), some families of mollusks [3] (mainly Conidae) and arthropods [4] (scorpions, pseudoscorpions, spiders, centipedes, ticks and hymenoptera insects, like bees, ants and wasps). Arthropods inject their venom through fangs (spiders and centipedes) or stingers (scorpions and pungent insects). The sting, in some insects, such as bees and wasps, is a modified egg-laying device, called ovipositor. Some caterpillars have venom defense glands associated to specialized bristles in the body known as urticating hairs, which can be lethal to humans (such as the moth *Lonomia*) [5]. Bees use an acidic poison (apitoxin), which causes pain in those bitten, to defend their hives and food stocks [6]. Wasps, on the other hand, use its venom to only paralyze the prey [7]. In this way, the pray can be stored alive in food chambers for the young. The ant *Polyrhachis dives* produces a poison that is applied topically on the victim for pathogen sterilization [8]. There are many other venomous and poisonous invertebrates, including jellyfish [9], bugs [10] and snails [11-13]. The sea wasp (*Chironex fleckeri*), also called box jellyfish, has about 500,000 cnidocytes in each tentacle, containing nematocysts, a harpoon-shaped mechanism that injects an extremely potent venom into the victim, which causes severe physical and psychological symptoms known as Irukandji syndrome. In many cases, this inoculation leads to death of the victim, that is why sea wasp is popularly known as "the world's most venomous creature" [14].

Loxoscelism is a condition produced by the bite of spiders from the genus *Loxosceles*, and is the only proven cause of necrosis in humans of arachnological origin [15]. *Loxosceles* spiders can be found worldwide. However, their distribution is heavily concentrated at the Western Hemisphere, particularly at the Americas, with more evidence in the tropics. In urban areas of South America, the presence of this type of spider is so evident that loxoscelism is considered a public health problem. Although *Loxosceles* bite is usually mild, it may ulcerate or cause more serious dermonecrotic injurie and even systemic reactions. This injurie is mainly due to the presence of the enzyme sphingomyelinase D in spider venom. Because the great number of diseases which mimic the loxoscelism symptoms, it is frequently misdiagnosed by physicians [16]. Although there is no known fully effective therapy for loxoscelism, research about potential antivenoms and vaccines has been exhaustive, presently also using the peptidomic approach [17], and many palliative therapies are reported in literature [15, 18].

Among vertebrates, only few reptiles (snakes and lizards) have developed the machinery for venom production [9], although some fish [9, 19], amphibians [20] and mammals (platypus for example) [21] have venom glands. The best known venomous reptiles are the snakes, which normally inject venom into their prey through hollow fangs. The snake venom is produced by mandibular glands located below the eyes and is inoculated into the victim through tubular or channeled fangs. Snakes use their venom mainly for hunting, although they can also use it for defense. A snakebite can cause a variety of symptoms including pain, swelling, tissue damage, decreased blood pressure, seizures, bleeding, respiratory paralysis, kidney failure and coma, and may, in severe cases, cause the patient death. These symptoms will vary depending on snake specie. Snakebite is an important medical emergency in many parts of the world, particularly in tropical and subtropical regions. According to World Health Organization (WHO), the incidence of snakebite reaches 5 million per year, causing 2.5 million envenomations and 125,000 deaths [22]. About 80% of envenomation deaths worldwide are caused by snakebite, followed by scorpion bite, which causes 15% [23]. Most affected are healthy people, such as children and agricultural populations, usually in poor resources areas, away from health centers in low-income countries in Africa, Asia and Latin America. As a result, WHO declared snakebite as a health crisis and a neglected tropical disease.

In addition to snakes, there are other venomous reptiles, such as the beaded lizard (*Heloderma horridum*), the Gila monster (*Heloderma suspectum*) and other species of lizards [9]. The composition of the Komodo dragon (*Varanus komodoensis*) venom is as complex as snake venoms [24]. Because of recent studies of venom glands in squamata and analysis of nuclear protein-coding genes, a new hypothetical clade, Toxicofera, is being proposed [25]. This clade would include all poisonous Squamata: suborders Serpentes (snakes) and Iguania (iguanas, agamid lizards, chameleons, etc.) and the infraorder Anguimorpha, represented by the families Varanidae (monitor lizards), Anguidae (alligator lizards, glass lizards, etc.) and Helodermatidae (Gila monster and beaded lizard).

Venoms can also be found in some fish, such as cartilaginous (rays, sharks and chimaeras) and teleostean, including monognathus eel-like fishes, catfishes, rockfishes, waspfishes, scorpionfishes, lionfishes, goatfishes, rabbitfishes, spiderfishes, surgeonfishes, gurnards, scats, stargazers, weever, swarmfish, etc. [9, 19]. Another venomous fish, the doctor fish, also know as "reddish log sucker", is used by some spas to feed the affected and dead areas of the skin of psoriasis patients, leaving the healthy skin to grow. There are venomous mammals, including solenodons, shrews, slow loris and the male platypus [21]. There are few poisonous amphibian species [20]. Some salamanders can expel venom through a rib of a sharp edge. There are even reports of venomous dinosaurs [26]. *Sinornithosaurus*, a genus of Dromaeosauridae dinosaur with feathers, may have had a venomous bite. But this theory is still controversial. The coelophysoid dinosaur *Dilophosaurus* is commonly portrayed in popular culture as being poisonous, but this superstition is not considered likely by the scientific community.

Until recently, the work in toxinology involved prospecting highly toxic or lethal toxins in animal venoms that could explain the symptoms observed clinically. Typically, such an approach involved the isolation and structural characterization of the molecule which causes an specific adverse effect observed when a person is envenomed. However, small molecules with micro-effects that were not easily observed were neglected or poorly studied. This situation changed in recent years with the improvement in sensitivity, resolution and accuracy of mass spectrometry and other techniques used in proteomic toxinology. With the adven of these new technologies, small peptides from animal venoms with unexplored biological activities started to be studied systematically, emerging, then, this new area of knowledge and scientific research called peptidomics. These molecules are potential candidates for new drugs or compounds with significant therapeutic actions.

2. Chemical composition and strategic importance of venoms

Over 5000 years ago, the Mesopotamians used a cane with a serpent as an emblem of Ningizzida, the god of fertility, marriage and pests. In Christianity, the serpent has always been associated with evil because of the biblical allegory of Adam and Eve. There is also a biblical story in which Moses erected a post with a brazen serpent to release his people from the plague of snakes. Throughout the development of Christianity, this symbol was transformed and the post became a tau.

But not always and not in all cultures, serpents were associated with evil. Many people believed in the cure power of serpents, often associated with its venom. Indeed, the medicinal value of animal venoms has been known since Antiquity. The medicinal use of bee venom, apitoxin, is reported in ancient Egypt and in Europe and Asia history. Charlemagne and Ivan *the Terrible*, for example, would have used apitoxin to treat common diseases. The medical uses of scorpion and snake venoms are well documented in Chinese pharmacopoeia. In an Islamic traditional tale, Muhammed is sick and, in the face of no known cure, it allows the use of snake venom as a last resource.

To Greeks and Romans, the serpent was a symbol of cure because periodically abandons its old skin and seemingly reborn, in the same way that doctors remove the disease of the body and rejuvenate the men, and also because the serpent was a symbol of concentrated attention, which was required to the curers. However, the association of serpents with cure may also be related to its venom, represented symbolically by herbs in the Greek-Roman mythology of Aesculapius, the god of medicine and cure. Called to assist Glaucus, who had been killed by lightning, Aesculapius saw a snake enter the room where he was, and killed it with his staff. Soon, a second serpent entered the room carrying herbs in its mouth, which deposited at the mouth of another dead serpent, making it back to life. Watching this scene, Aesculapius decided to put the herbs into the Glaucus mouth, who also raised from dead. Since then, Aesculapius turned the serpent your pet guardianship. His staff with a coiled serpent became the symbol of modern medicine in a large number of countries and is present even in the banner of the World Health Organization (WHO).

However, despite the healing power of animal venoms be known for a long time, the systematic investigation of venom components as natural sources for the generation of pharmaceuticals was only performed over the past decades, after a peptide that potencialize bradykinin action was isolated from the venom of the Brazilian snake *Bothrops jararaca* [27]. This led to the development, in the 1950s, of the first commercial drug based on angiotensin I converting enzyme (ACE)-inhibitor (trade name captopril®), for the treatment of arterial hypertension and heart failure [28]. Prialt® (ziconotide) is another example of synthetic drug successfully isolated from an animal venom [11]. This is a synthetic non-opioid peptide, non-NSAID, non-local anesthetic calcium channel blocker, isolated from the secretions of the cone snail *Conus magus*. Prialt® is used for the alleviation of chronic intractable pain and is administered directly into the spinal cord, due to deep side effects or lack of efficacy when it is administered by the more common routes such as orally or intravenously.

The evolution of the venom secretion apparatus in animals is indeed an impressive biological achievement at the evolutionary point of view. Since venoms components result of biochemical and pharmacological refinement over a long period in evolutionary scale, they have been tuned for optimum activity by the natural evolution. Thus, nature has already prospected huge combinatorial libraries of potential therapeutic drugs. The biochemical evolution of proteins from salivary fluids or venom exocrine glands is remarkable, especially when one considers the highly specialized functions of these proteins and its high specificity with respect to the target molecule.

Several classes of organic molecules have been described in venoms, such as alkaloids and acylpolyamines. However, the main constituents are indeed polypeptides. Venoms of cone snail and arthropods, such as spiders, scorpions and insects, to a lesser extent, seem to be mainly peptidic, while snakes produce protein rich venoms. Snake venoms contain a variety of proteases, which hydrolyze peptide bonds of proteins, nucleases, which hydrolyze phosphodiester bonds of DNA, and neurotoxins, which disable signaling in the nervous system. The brown spider venom contains a variety of toxins, the most important of which is the tissue destruction agent sphingomyelinase D, present in the venom of all species of *Loxosceles* in different concentrations [29]. Only another spider genus (*Sicarius*) and several pathogenic bacteria are known to produce this enzyme.

Some venoms comprise several hundreds of components, which further expands its potential as a source of new medicines. Many components of venoms affect the nervous system and modulate the generation and propagation of action potentials, acting on multiple molecular sites, which include central and peripheral neurons, axons, synapses and neuromuscular junctions [30]. Many of these target receptors play important physiological roles or are associated with specific diseases. Therefore, the components of animal venoms are important biological tools for studying these receptors, and the discovery of molecules in venoms with selective activity for these receptors represents a very attractive approach to the search for new drugs. The venom components may therefore be probed for the development of new therapies for pain management [31], new anti-arrhythmic [32], anticonvulsant [33] or anxiolytic drugs [34], new antimicrobial agents [35-37] or pesticides [38, 39], etc. Even a substance that causes priapism has been isolated from the venom of a Brazilian spider [40], becoming a potential drug candidate to attend erectile dysfunction.

Another reason to study the composition of animal venoms is trying to seek more effective prophylaxis for envenomings. Doctors treat victims of venomous sting with serum, which is produced by injecting into an animal, such as sheep, horse, goat or rabbit, a small amount of specific venom. The animal's immune system responds to the target dose, producing antibodies to active molecules of the venom. These antibodies can then be isolated from the animal's blood and used in envenoming treatment in other animals, including humans

However, this treatment can be effectively used only a limited number of times for a particular person, since that person will develop antibodies to neutralize the exogenous animal's antibodies used to produce the antiserum (antibodies antiantibodies). Even if that person does not suffer a severe allergic reaction to the antiserum, his own immune system can destroy the antiserum even before the antiserum destroys the venom toxins. Most people will never need an antiserum treatment throughout their lives. However, others, who work or live in risk areas habited by snakes or other venomous animals, such as agricultural areas for example, need that this treatment is available in public health network.

Some treatments are done not with antiserum, but herbal. *Aristolochia rugosa* and *Aristolochia trilobata*, or angelic, are medicinal plants used in Western India and in Central and South America against snake and scorpion bites [41]. Aristolochic acid, produced by those plants, inhibits inflammation induced by immune complexes and non-immunological agents (carrageenan or croton oil). It also inhibits the activity of phospholipases present in snake venoms (PLA2], forming a 1:1 complex with the enzyme. Phospholipases play an important role in the reactions cascade that lead to inflammatory response and pain. Therefore, its inhibition may reduce problems of scorpionism, snakebite and loxoscelism.

3. Proteomic and peptidomic analysis: A new approach to study venoms

Proteins are very large molecules formed by amino acids chains linked together as a polymer. Although biological systems uses only 20 amino acids to build their proteins, the different possible combinations among them is virtually infinite, resulting in tens of thousands different proteins, each one with a unique sequence, genetically defined, which determines its specific form and biological function. Furthermore, each protein may undergo a variety of post-translational modifications, which diversifies even more its form and function. Proteins are the main constituents of protoplasm of all cells. As the major components of cells metabolic pathways, proteins have vital functions in organism, such as: catalyze biochemical reactions (eg. enzymes), transmit messages (eg. neurotransmitters), regulate cellular reproduction, influence growth and development of various tissues (eg. trophic factors), carry oxygen in the blood (eg. hemoglobin), defend the body against diseases (eg. antibodies), among countless other achievements. There is no metabolic reaction in which the participation of at least one protein is dispensable.

The term "proteome" is derived from the junction of the word "PROTEin" with the word "genOME" and refers to the set of proteins expressed starting from a genome, i.e., all the proteins produced by an organism. Indeed, the word proteome is often be more related to the set of proteins expressed in a specific organ, or biological fluid, or cell, in a given state (eg. diseased cell). The proteome is therefore the complete complement of a genome, including the "makeup" that proteins receives after being synthesized, i.e., the post-translational modifications, all of them absolutely relevant for that proteins perform their biological function. The proteome of a cell or fluid varies with time and conditions under which the organism is subjected. The human body, for example, can contain more than 2 million different proteins, each one exerting a distinct role. Unlike the genome, which is relatively static, the proteome is constantly changing in response to tens of thousands of intra and extracellular environmental signals. The proteome varies with the nature of each tissue or organ, the cell development stage, the stress conditions to which the organism is subjected, the organism health state, the effects of drug treatment, etc. As such, the proteome is often defined as the proteins present in a sample (tissue, organism, cell culture, biological fluid, etc.) at a given point in time.

The term proteomics consists of comprehensive and systematic study of all proteins present in a given cell state, which was made possible by the huge development of mass spectrometry techniques over the past two decades. Proteomics and genomics run parallel and are interdependent. Genomics without proteomics is only an "alphabet soup", because it can only make inferences about their products (proteins). Moreover, proteomics requires genomics to identify the proteins expressed in a particular cell state. Briefly, genomics provides a static information of the various ways in which a cell may use its proteins, while proteomics gives a dynamic panorama of molecular diversity, showing not only which proteins are more or less expressed (or is not even expressed), but also how these proteins were modified and how these modifications affect its role in the cell theater.

Proteomic technologies can play an important role in new drugs discovery, new diagnostics and molecular medicine, because it is the connection between genes, proteins and diseases. For example, the discovery of defective proteins that cause specific diseases can help develop new drugs that either alter the shape of a defective protein or mimic its action. Most of the most popular drugs today either have proteinaceous nature or have a protein target. Through proteomics, one can create "custom" drugs, i.e., drugs specially designed for specific individuals. Such drugs are supposed to be more effective and cause fewer side effects. Another field to which proteomic studies can contribute is the biomarkers discovery for specific diseases, whose overexpression (or depletion) would indicate, quite early, the disease development. For example, serum levels of prostate specific antigen (PSA) is commonly used in the diagnosis of prostate cancer in men, which makes PSA a biomarker for cancer. Unfortunately, however, the diagnosis based on a single protein biomarker is not very reliable. Proteomics may help scientists to develop diagnostic tests that simultaneously analyze the expression of multiple proteins in order to improve the specificity and sensitivity of these tests.

Over time, new study areas with the suffix "omics" have emerged, such as metabolomics, lipidomics, carbohydratomics, degradomics etc. The term venomics did not slow to appear, and today it is defined as the study of all components (protean or not protean) of a venom. The word peptidomics has also been proposed to set the study of the peptides (instead of proteins) of a cell type or a biological fluid, such as venom. According to Ivanov and Yatskin [42]: "structure and biologic function of the entire multitude of peptides circulating in living organisms, their organs, tissues, cells and fluids comprises the scope of peptidomics". For these authors, "these two multitudes of polypeptides (proteins and peptides) play a dominant role in the functioning of any cellular system, tissue or organ. They are intimately connected with each other and exist in equilibrium as an essential part of homeostasis (i.e., the normal state of any living organism and the basis of life itself)".

Peptidomic analysis has been proposed by several authors [43-55] as a way to access information relevant to clinical diagnosis and/or to monitor the patient biochemical profile during the therapy. The growing interest in peptidomic analysis led some scientists to develop new analytical technologies to improve peptidomic analysis, such as: use of capillary electrophoresis to separate the peptides [46]; use of size exclusion chromatography as a pre-fractionation step [53, 56]; new technologies and methods for sample pretreatment [57], such as methods for isolation rare amino acid-containing peptides, terminal peptides, PTM peptides and endogenous peptides, automated sample pretreatment technologies (automated sample injection and on-line digestion) [58]; development of a new target plate for MALDI-MS for one step electric transfer of analytes from a 1-dimensional electrophoresis gel directly to the target plate [59, 60]; etc. In recent years, in the face of the remarkable development on nanotechnology, many researchers have produced different kind of nanoparticles, such as mesoporous silica nanoparticles [50, 51, 61, 62] and carbon nanotubes [52, 63], for selective peptide extraction (and, hence, its enrichment) from biological fluids for therapeutic purposes (clinical diagnosis and/or novel biomarker discovery).

In the case of animal venoms, however, peptidomics is a highly interesting area for different reasons, since most of the biologically active components of pharmacological interest are of peptidic nature [64]. For example, Biass and co-workers [12] studied the venom peptidomic profile of the cone snail-hunting fish, *Conus consors*, through approaches involving different sample preparation protocols and analysis by mass spectrometry. The cone snail was quoted in the television series *Animal Planet: The Most Extreme*, because it can quickly shoot a harpoon filled with deadly toxins. The conidia (Conidae) constitute a family of several shells divided into subfamilies. It is estimated that this genus produce more than 70,000 different pharmacologically active components, most of peptidic nature, whereas interspecies variations. It is a rich library of neuropharmacology and combinatorial chemistry. Precisely for this reason, the 6th Framework Programme of the European Union funded with € 10.7 million the international project CONCO involving 20 partners and 13 countries [65], whose objective is to explore new molecules therapeutically relevant produced by venomous marine cone snails.

4. The tools to peptidomic analysis

Mass spectrometry is an analytical tool that has evolved dramatically over the past 20 years in terms of sensitivity, resolving power and versatility, and is currently one of the main tools for studying the molecular components of biological systems, including venoms. The development of techniques such as electrospray ionization (ESI) and matrix-assisted laser desorption ionization (MALDI) was essential for allowing polypeptides be analyzed by mass spectrometry. Hyphenation of separation techniques such as high performance liquid chromatography (HPLC) with mass spectrometry was also decisive for this progress. As a consequence, the highly combinatorial nature of venom components and their underlying pharmacologic complexity have been progressively revealed by mass spectrometry. Currently, major challenges remain on samples complexity, lack of biological material and databases absence to peptide and protein identification based on sequence information.

Peptidomic analysis of a sample will consist of essentially four steps: (I) peptides extraction from the sample; (II) separation of these peptides — including their prior separation from other polypetidic components of the sample, i.e., proteins, defined as the protean components with molecular weight above 10 kDa —; (III) peptides detection — which is commonly performed by mass spectrometry —, (IV) and finally identification of the peptides — which usually involves fragmentation of those peptides in a tandem mass spectrometer (MS/MS).

With respect to peptide sequencing for identification purposes, the technique traditionally used is Edman degradation-based sequencing [66, 67]. But nowadays this kind of sequencing is increasingly being replaced by sequencing techniques based on mass spectrometry [68, 69]. This is due to the fact that mass spectrometry is much more rapid and sensitive than Edman sequencing and prenscinde of prior separation of the peptides, which means that peptides can be sucessfully analyzed and sequenced by mass spectrometry from a complex peptide matrix, which is impossible by Edman sequencing. This is only possible because the peptide of interest is selected (i.e., separated from others) in the first mass spectrometer. Then, this parent ion is fragmented in a collision chamber and the daughter ions are analyzed in a second mass spectrometer (MS/MS). Figure 3 gives an example of peptide *de novo* sequencing by tandem mass spectrometry. For more details about this kind of polypeptide sequencing, see reference [69].

In proteomics, the most widely used technique to separate protean components of a sample is the two-dimensional polyacrylamide gel electrophoresis (2D-PAGE). In peptidomics, however, techniques based on liquid chromatography coupled to mass spectrometry (LC-MS) appear to be more popular, since peptides are not well resolved by electrophoresis [70]. Despite this, capillary electrophoresis has also been used successfully in peptidomic analysis, mainly to analyze biological fluids for clinical applications, such as disease diagnosis and response to therapy [46].

As an example, Valente and co-workers [71] ran a two-dimensional gel from the venom of *Bothrops insularis*, an endemic snake specie in Queimada Grande Island, Brazil. The result is shown in Figure 1. This is an example of venomics, i.e., the study of all protean components of a venom. Using the proteomic approach, the authors detected 494 spots in the gel using an image analysis software, from which 69 proteins were identified by current identification techniques, using mass spectrometry and heavy bioinformatics to interpret the mass spectra and also to make a comparative search of protein sequences deposited in databases. The identified proteins include metalloproteinases, serine proteinases, phospholipases A2, lectins, growth factors, L-amino acid oxidases, the developmental protein G10, a disintegrin, a nuclear protein of the BUD31 family, and putative novel bradykinin-potentiating peptides. In the same study, the authors also performed a peptidomic analysis of the venom, by direct analysis of the crude venom by MALDI-TOF-TOF and LC-ESI-Q-TOF. Many new peptides were partially or completely sequenced by both MALDI-MS/MS and LC-ESI-MS/MS. Using the proteomic approach associated with peptidomic analysis, the authors could speculate about the existence of posttranslational modifications and a proteolytic processing of precursor molecules which could lead to diverse multifunctional proteins.

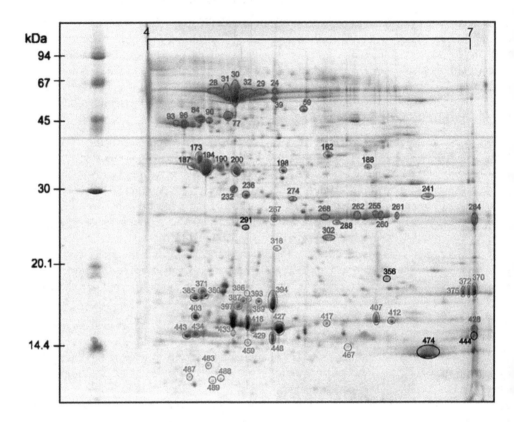

Figure 1. Proteomic profile of *Bothrops insularis* venom: 2D-PAGE reference map (copyed from reference [71]).

Liao and co-workers [72] also applied proteomic and peptidomic approaches together to analyze the venom of *Chilobrachys jingzhao* (a type of tarantula; one of the most venomous spiders in southern China). They developed a protocol which consists in run a gel filtration of the crude venom and then divide the fractions in two parts. The fraction containing protean components with molecular mass above 10 kDa they underwent proteomic analysis, which consisted of 2-DE, in gel trypsin digestion, MALDI-TOF-TOF and ESI-Q-TOF analysis of the spots, protein identification by PMF, *de novo* sequencing of the peptides, and protein identification by MS BLAST sequence similarity search. The fraction containing protein components with molecular weight below 10 kDa was used for peptidomic analysis, consisting in separation of the peptides by ion-exchange HPLC followed by reverse phase HPLC, MALDI-TOF analysis of the chromatographic fractions, Edman peptide sequencing, and peptide identification by MS BLAST. The authors reported that peptides were the predominant components [69%) of the dry crude venom, while proteins accounted only for 6%. Nonprotean components (low MW inorganic and organic molecules, such as polyamines, salts, free acids, glucose, etc.) complete the remaining 25% of the crude venom.

Another good example of peptidomic analysis was presented by Rates and co-workers [73], who studied the *Tityus serrulatus* (a specie of scorpion whose venom has been most extensively studied) venom peptide diversity. In this work, the authors fracionated the venom by gel filtration followed by reverse phase chromatography of each fraction obtained in the first separation. The results are shown in Figure 2. Then, the chromatographic fractions were analyzed by MALDI-TOF-TOF. The peptides were sequenced using *de novo* methodology (Figure 3) and the sequences obtained were compared with protein databases in sequence similarity searches. The authors also reported the finding of novel peptides without sequence similarities to other known molecules.

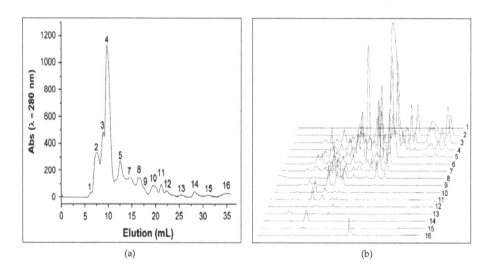

Figure 2. *Tityus serrulatus* venom fractionation through gel filtration (A) and re-chromatography of each fraction by reverse phase chromatography (B) (copyed from reference [73]).

One of the biggest difficulties currently encountered by researchers working with peptidomic analysis of animal venoms is that organisms with unsequenced genomes, including venomous animals, still represent the overwhelming majority of species in the biosphere. Fortunately, Andrej Shevchenko, from Max Planck Institute of Molecular Cell Biology and Genetics, at Dresden, Germany, paved the way for homology-driven proteomic approaches to explore proteomes of organisms with unsequenced genomes [74-76]. Through this new methodology, the search against sequences databases is made not by the exact sequence, but by sequence similarity to other protein sequences deposited in the database. This new approach does not fully solve the problem, but allows peptides to be positively identified in peptidomic experiments through cross-species identification. Wang and colleagues [77] developed an alternative strategy to circumvent the problem of absence of systematic online database information, and used this technique to analyze the peptidome of amphibian skin secretions. Although amphibian skin secretion is not exactly a venom, it is still a biological model also very promising for the search of new pharmacologically active substances. First, the authors deduced all of putative bioactive peptide sequences by shotgun cloning the cDNAs encoding peptide precursors. Then, they separated the entire peptidome by UPLC/MS/MS, and confirmed those sequences deduced before by *de novo* MS/MS sequencing.

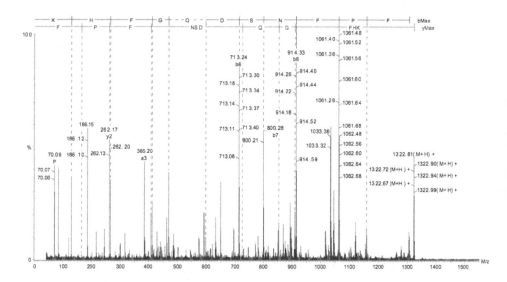

Figure 3. MS/MS spectra interpretation (*de novo* sequencing) for peptide NH$_2$-FPFNSD(K/Q)GFH(K/Q)-CO$_2$H (copyed from reference [73]). The K/Q denotes a doubt about the possibility of being lysine or glutamine, as these two amino acids are isobaric. However, as trypsin cleaves on C-terminal sides of arginines and lysines, it is likely that the middle amino acid is glutamine and the last one is lysine.

5. Concluding remarks

Animal venoms are true "cocktails" of substances normally harmful, but that can be explored with intelligence for medical use. Many authors even use the word "cornucopia" to define a venom. The cornucopia — junction of the Latin *"cornu"* (horn) with *"copiae"* (strength) -, also called "horn of plenty", is a symbol of nourishment and abundance in classical mythology, usually represented by a large horn-shaped container overflowing with products such as flowers, dried fruit, other foodstuffs, and other types of wealthiness. Nowadays it is particularly associated with the Thanksgiving holiday. In toxinology, cornucopia represents the chemical wealth of animal venoms, where one can find thousands of such substances with interesting biological effects that can be explored as candidates for future pharmaceutical products.

Much (perhaps the largest) of this pool of substances is of peptidic nature, i.e., polypeptides with molecular weight below 10 kDa. These are biologically active peptides with diverse functions, ranging from heart hypotensors to erectile dysfunction controllers. Thus, the peptidomic analysis of animal venoms is an emerging and promising area of science, and can be considered a frontier area as it includes researchers from toxinology, proteomics, pharmacology, therapeutics, drug discovery, peptide chemistry, analytical chemistry, etc.

In this chapter we tried to show the importance of animal venoms for molecular toxinology and its potential use for biomedical applications. We also sought to demonstrate the recent advent and rapid growth of peptidomic analysis as the main tool to explore the molecular features of these venoms, not only to produce more efficient antisera against venomous bites but also and mainly to characterize the components of peptide nature in search for new products of pharmacological interest. Although this new science is still in its early stages of development, it is already very mature. This is a science field that has enough potential to grow and provide creative solutions to problems that affect human health. Hopefully more and more researchers become interested on this topic. Medicine has much to gain from it.

Author details

Ricardo Bastos Cunha

Bioanalytical Chemistry Laboratory, Division of Analytical Chemistry, Institute of Chemistry, University of Brasília, Brazil

References

[1] Russell FE. Venomous animal injuries. Current problems in pediatrics. 1973;3(9):1-47. Epub 1973/07/01.

[2] Bedry R, de Haro L. [Venomous and poisonous animals. V. Envenomations by venomous marine invertebrates]. Medecine tropicale : revue du Corps de sante colonial. 2007;67(3):223-31. Epub 2007/09/06. Envenimations ou intoxications par les animaux venimeux ou veneneux. V--invertebres marins venimeux.

[3] Hermitte LC. Venomous marine molluscs of the genus Conus. Transactions of the Royal Society of Tropical Medicine and Hygiene. 1946;39:485-512. Epub 1946/06/01.

[4] Ghosh SM. Injuries by Venomous Arthropods. Bulletin of the Calcutta School of Tropical Medicine. 1965;13:30-3. Epub 1965/01/01.

[5] Veiga AB, Blochtein B, Guimaraes JA. Structures involved in production, secretion and injection of the venom produced by the caterpillar Lonomia obliqua (Lepidoptera, Saturniidae). Toxicon : official journal of the International Society on Toxinology. 2001;39(9):1343-51. Epub 2001/06/01.

[6]   Tannenberg J, Kosseff A. Fatal anaphylactic shock due to a bee sting in the finger. Proceedings of the New York State Association of Public Health Laboratories. 1945;25(2):33. Epub 1945/01/01.

[7]   Seward EH. Wasp venoms and anaesthesia. Proceedings of the Royal Society of Medicine. 1954;47(12):1032-4. Epub 1954/12/01.

[8]   Blum MS, Walker JR, Callahan PS, Novak AF. Chemical, insecticidal and antibiotic properties of fire ant venom. Science. 1958;128(3319):306-7. Epub 1958/08/08.

[9]   Warrell DA. Venomous bites, stings, and poisoning. Infectious disease clinics of North America. 2012;26(2):207-23. Epub 2012/05/29.

[10] Viswanathan M, Srinivasan K. Poisoning by bug poison. A preliminary study. Journal of the Indian Medical Association. 1962;39:345-9. Epub 1962/10/01.

[11] McIntosh M, Cruz LJ, Hunkapiller MW, Gray WR, Olivera BM. Isolation and structure of a peptide toxin from the marine snail Conus magus. Archives of biochemistry and biophysics. 1982;218(1):329-34. Epub 1982/10/01.

[12] Biass D, Dutertre S, Gerbault A, Menou JL, Offord R, Favreau P, et al. Comparative proteomic study of the venom of the piscivorous cone snail Conus consors. Journal of proteomics. 2009;72(2):210-8. Epub 2009/05/22.

[13] Eugster PJ, Biass D, Guillarme D, Favreau P, Stocklin R, Wolfender JL. Peak capacity optimisation for high resolution peptide profiling in complex mixtures by liquid chromatography coupled to time-of-flight mass spectrometry: Application to the Conus consors cone snail venom. Journal of chromatography A. 2012. Epub 2012/06/05.

[14] Pearn J, Fenner P. The Jellyfish hunter--Jack Barnes: a pioneer medical toxinologist in Australia. Toxicon : official journal of the International Society on Toxinology. 2006;48(7):762-7. Epub 2006/10/31.

[15] Swanson DL, Vetter RS. Loxoscelism. Clinics in dermatology. 2006;24(3):213-21. Epub 2006/05/23.

[16] Bennett RG, Vetter RS. An approach to spider bites. Erroneous attribution of dermonecrotic lesions to brown recluse or hobo spider bites in Canada. Canadian family physician Medecin de famille canadien. 2004;50:1098-101. Epub 2004/10/01.

[17] Guimarães AB. Análise peptidômica comparativa das peçonhas de duas espécies de aranhas marrom: Loxosceles laeta e Loxosceles intermedia [dissertation]. http://capesdw.capes.gov.br/capesdw/resumo.html?idtese=20092753001010005P5: University of Brasilia; 2009.

[18] da Silva PH, da Silveira RB, Appel MH, Mangili OC, Gremski W, Veiga SS. Brown spiders and loxoscelism. Toxicon : official journal of the International Society on Toxinology. 2004;44(7):693-709. Epub 2004/10/27.

[19] Maretic Z. [Experience with venomous fish bites]. Acta tropica. 1957;14(2):157-61. Epub 1957/01/01. Erfahrungen mit Stichen von Giftfischen.

[20] Chippaux JP, Goyffon M. [Venomous and poisonous animals--I. Overview]. Medecine tropicale : revue du Corps de sante colonial. 2006;66(3):215-20. Epub 2006/08/24. Envenimations et intoxications par les animaux venimeux ou veneneux--I. General-its.

[21] Whittington CM, Koh JM, Warren WC, Papenfuss AT, Torres AM, Kuchel PW, et al. Understanding and utilising mammalian venom via a platypus venom transcriptome. Journal of proteomics. 2009;72(2):155-64. Epub 2009/01/21.

[22] Girish KS, Kemparaju K. Overlooked issues of snakebite management: time for strategic approach. Current topics in medicinal chemistry. 2011;11(20):2494-508. Epub 2011/06/28.

[23] Calvete JJ, Sanz L, Angulo Y, Lomonte B, Gutierrez JM. Venoms, venomics, antivenomics. FEBS Lett. 2009;583(11):1736-43. Epub 2009/03/24.

[24] Fry BG, Wroe S, Teeuwisse W, van Osch MJ, Moreno K, Ingle J, et al. A central role for venom in predation by Varanus komodoensis (Komodo Dragon) and the extinct giant Varanus (Megalania) priscus. Proceedings of the National Academy of Sciences of the United States of America. 2009;106(22):8969-74. Epub 2009/05/20.

[25] Vidal N, Hedges SB. The phylogeny of squamate reptiles (lizards, snakes, and amphisbaenians) inferred from nine nuclear protein-coding genes. Comptes rendus biologies. 2005;328(10-11):1000-8. Epub 2005/11/16.

[26] Gong E, Martin LD, Burnham DA, Falk AR. The birdlike raptor Sinornithosaurus was venomous. Proceedings of the National Academy of Sciences of the United States of America. 2010;107(2):766-8. Epub 2010/01/19.

[27] Rocha ESM, Beraldo WT, Rosenfeld G. Bradykinin, a hypotensive and smooth muscle stimulating factor released from plasma globulin by snake venoms and by trypsin. The American journal of physiology. 1949;156(2):261-73. Epub 1949/02/01.

[28] Fernandez JH, Neshich G, Camargo AC. Using bradykinin-potentiating peptide structures to develop new antihypertensive drugs. Genetics and molecular research : GMR. 2004;3(4):554-63. Epub 2005/02/03.

[29] Merchant ML, Hinton JF, Geren CR. Sphingomyelinase D activity of brown recluse spider (Loxosceles reclusa) venom as studied by 31P-NMR: effects on the time-course of sphingomyelin hydrolysis. Toxicon : official journal of the International Society on Toxinology. 1998;36(3):537-45. Epub 1998/06/24.

[30] Pungercar J, Krizaj I. Understanding the molecular mechanism underlying the presynaptic toxicity of secreted phospholipases A2. Toxicon : official journal of the International Society on Toxinology. 2007;50(7):871-92. Epub 2007/10/02.

[31] Reid PF. Cobra venom: A review of the old alternative to opiate analgesics. Alternative therapies in health and medicine. 2011;17(1):58-71. Epub 2011/05/28.

[32] Ma H, Zhou J, Jiang J, Duan J, Xu H, Tang Y, et al. The novel antidote Bezoar Bovis prevents the cardiotoxicity of Toad (Bufo bufo gargarizans Canto) Venom in mice. Experimental and toxicologic pathology : official journal of the Gesellschaft fur Toxikologische Pathologie. 2012;64(5):417-23. Epub 2010/11/19.

[33] Fachim HA, Cunha AO, Pereira AC, Beleboni RO, Gobbo-Neto L, Lopes NP, et al. Neurobiological activity of Parawixin 10, a novel anticonvulsant compound isolated from Parawixia bistriata spider venom (Araneidae: Araneae). Epilepsy & behavior : E&B. 2011;22(2):158-64. Epub 2011/07/19.

[34] Cesar-Tognoli LM, Salamoni SD, Tavares AA, Elias CF, Costa JC, Bittencourt JC, et al. Effects of Spider Venom Toxin PWTX-I (6-Hydroxytrypargine) on the Central Nervous System of Rats. Toxins. 2011;3(2):142-62. Epub 2011/11/10.

[35] Samy RP, Stiles BG, Gopalakrishnakone P, Chow VT. Antimicrobial proteins from snake venoms: direct bacterial damage and activation of innate immunity against Staphylococcus aureus skin infection. Current medicinal chemistry. 2011;18(33): 5104-13. Epub 2011/11/05.

[36] Azevedo Calderon L, Silva Ade A, Ciancaglini P, Stabeli RG. Antimicrobial peptides from Phyllomedusa frogs: from biomolecular diversity to potential nanotechnologic medical applications. Amino acids. 2011;40(1):29-49. Epub 2010/06/08.

[37] Remijsen Q, Verdonck F, Willems J. Parabutoporin, a cationic amphipathic peptide from scorpion venom: much more than an antibiotic. Toxicon : official journal of the International Society on Toxinology. 2010;55(2-3):180-5. Epub 2009/10/31.

[38] Bosmans F, Tytgat J. Sea anemone venom as a source of insecticidal peptides acting on voltage-gated Na+ channels. Toxicon : official journal of the International Society on Toxinology. 2007;49(4):550-60. Epub 2007/01/17.

[39] Rohou A, Nield J, Ushkaryov YA. Insecticidal toxins from black widow spider venom. Toxicon : official journal of the International Society on Toxinology. 2007;49(4): 531-49. Epub 2007/01/11.

[40] Nunes KP, Costa-Goncalves A, Lanza LF, Cortes SF, Cordeiro MN, Richardson M, et al. Tx2-6 toxin of the Phoneutria nigriventer spider potentiates rat erectile function. Toxicon : official journal of the International Society on Toxinology. 2008;51(7): 1197-206. Epub 2008/04/10.

[41] Martz W. Plants with a reputation against snakebite. Toxicon : official journal of the International Society on Toxinology. 1992;30(10):1131-42. Epub 1992/10/01.

[42] Ivanov VT, Yatskin ON. Peptidomics: a logical sequel to proteomics. Expert review of proteomics. 2005;2(4):463-73. Epub 2005/08/16.

[43] Tammen H, Schulte I, Hess R, Menzel C, Kellmann M, Mohring T, et al. Peptidomic analysis of human blood specimens: comparison between plasma specimens and serum by differential peptide display. Proteomics. 2005;5(13):3414-22. Epub 2005/07/23.

[44] Tammen H, Hess R, Schulte I, Kellmann M, Appel A, Budde P, et al. Prerequisites for peptidomic analysis of blood samples: II. Analysis of human plasma after oral glucose challenge -- a proof of concept. Combinatorial chemistry & high throughput screening. 2005;8(8):735-41. Epub 2006/02/09.

[45] Tammen H, Schulte I, Hess R, Menzel C, Kellmann M, Schulz-Knappe P. Prerequisites for peptidomic analysis of blood samples: I. Evaluation of blood specimen qualities and determination of technical performance characteristics. Combinatorial chemistry & high throughput screening. 2005;8(8):725-33. Epub 2006/02/09.

[46] Schiffer E, Mischak H, Novak J. High resolution proteome/peptidome analysis of body fluids by capillary electrophoresis coupled with MS. Proteomics. 2006;6(20): 5615-27. Epub 2006/09/23.

[47] Geho DH, Liotta LA, Petricoin EF, Zhao W, Araujo RP. The amplified peptidome: the new treasure chest of candidate biomarkers. Current opinion in chemical biology. 2006;10(1):50-5. Epub 2006/01/19.

[48] Traub F, Jost M, Hess R, Schorn K, Menzel C, Budde P, et al. Peptidomic analysis of breast cancer reveals a putative surrogate marker for estrogen receptor-negative carcinomas. Laboratory investigation; a journal of technical methods and pathology. 2006;86(3):246-53. Epub 2006/02/18.

[49] Mischak H, Julian BA, Novak J. High-resolution proteome/peptidome analysis of peptides and low-molecular-weight proteins in urine. Proteomics Clinical applications. 2007;1(8):792. Epub 2007/07/10.

[50] Tian R, Zhang H, Ye M, Jiang X, Hu L, Li X, et al. Selective extraction of peptides from human plasma by highly ordered mesoporous silica particles for peptidome analysis. Angew Chem Int Ed Engl. 2007;46(6):962-5. Epub 2006/12/14.

[51] Tian R, Ye M, Hu L, Li X, Zou H. Selective extraction of peptides in acidic human plasma by porous silica nanoparticles for peptidome analysis with 2-D LC-MS/MS. Journal of separation science. 2007;30(14):2204-9. Epub 2007/08/09.

[52] Li X, Xu S, Pan C, Zhou H, Jiang X, Zhang Y, et al. Enrichment of peptides from plasma for peptidome analysis using multiwalled carbon nanotubes. Journal of separation science. 2007;30(6):930-43. Epub 2007/06/01.

[53] Hu L, Li X, Jiang X, Zhou H, Jiang X, Kong L, et al. Comprehensive peptidome analysis of mouse livers by size exclusion chromatography prefractionation and nanoLC-MS/MS identification. Journal of proteome research. 2007;6(2):801-8. Epub 2007/02/03.

[54] Tammen H, Hess R, Rose H, Wienen W, Jost M. Peptidomic analysis of blood plasma after in vivo treatment with protease inhibitors--a proof of concept study. Peptides. 2008;29(12):2188-95. Epub 2008/09/23.

[55] Gelman JS, Sironi J, Castro LM, Ferro ES, Fricker LD. Peptidomic analysis of human cell lines. Journal of proteome research. 2011;10(4):1583-92. Epub 2011/01/06.

[56] Hu L, Ye M, Zou H. Peptidome analysis of mouse liver tissue by size exclusion chromatography prefractionation. Methods Mol Biol. 2010;615:207-16. Epub 2009/12/17.

[57] Beaudry F. Stability comparison between sample preparation procedures for mass spectrometry-based targeted or shotgun peptidomic analysis. Analytical biochemistry. 2010;407(2):290-2. Epub 2010/08/24.

[58] Jiang X, Ye M, Zou H. Technologies and methods for sample pretreatment in efficient proteome and peptidome analysis. Proteomics. 2008;8(4):686-705. Epub 2008/01/23.

[59] Tanaka K, Tsugawa N, Kim YO, Sanuki N, Takeda U, Lee LJ. A new rapid and comprehensive peptidome analysis by one-step direct transfer technology for 1-D electrophoresis/MALDI mass spectrometry. Biochemical and biophysical research communications. 2009;379(1):110-4. Epub 2008/12/17.

[60] Araki Y, Nonaka D, Tajima A, Maruyama M, Nitto T, Ishikawa H, et al. Quantitative peptidomic analysis by a newly developed one-step direct transfer technology without depletion of major blood proteins: its potential utility for monitoring of pathophysiological status in pregnancy-induced hypertension. Proteomics. 2011;11(13):2727-37. Epub 2011/06/02.

[61] Tian R, Ren L, Ma H, Li X, Hu L, Ye M, et al. Selective enrichment of endogenous peptides by chemically modified porous nanoparticles for peptidome analysis. Journal of chromatography A. 2009;1216(8):1270-8. Epub 2008/10/22.

[62] Hu L, Ye M, Zou H. Recent advances in mass spectrometry-based peptidome analysis. Expert review of proteomics. 2009;6(4):433-47. Epub 2009/08/18.

[63] Li F, Dever B, Zhang H, Li XF, Le XC. Mesoporous materials in peptidome analysis. Angew Chem Int Ed Engl. 2012;51(15):3518-9. Epub 2012/02/07.

[64] Pimenta AM, De Lima ME. Small peptides, big world: biotechnological potential in neglected bioactive peptides from arthropod venoms. Journal of peptide science : an official publication of the European Peptide Society. 2005;11(11):670-6. Epub 2005/08/17.

[65] ; Available from: http://www.toxinomics.org/conco_project.html.

[66] Edman P. A method for the determination of amino acid sequence in peptides. Archives of biochemistry. 1949;22(3):475. Epub 1949/07/01.

[67] Niall HD. Automated Edman degradation: the protein sequenator. Methods in enzymology. 1973;27:942-1010. Epub 1973/01/01.

[68] Roepstorff P, Fohlman J. Proposal for a common nomenclature for sequence ions in mass spectra of peptides. Biomedical mass spectrometry. 1984;11(11):601. Epub 1984/11/01.

[69] Steen H, Mann M. The ABC's (and XYZ's) of peptide sequencing. Nature reviews Molecular cell biology. 2004;5(9):699-711. Epub 2004/09/02.

[70] Songping L. Protocols for peptidomic analysis of spider venoms. Methods Mol Biol. 2010;615:75-85. Epub 2009/12/17.

[71] Valente RH, Guimaraes PR, Junqueira M, Neves-Ferreira AG, Soares MR, Chapeaurouge A, et al. Bothrops insularis venomics: a proteomic analysis supported by transcriptomic-generated sequence data. Journal of proteomics. 2009;72(2):241-55. Epub 2009/02/13.

[72] Liao Z, Cao J, Li S, Yan X, Hu W, He Q, et al. Proteomic and peptidomic analysis of the venom from Chinese tarantula Chilobrachys jingzhao. Proteomics. 2007;7(11): 1892-907. Epub 2007/05/04.

[73] Rates B, Ferraz KK, Borges MH, Richardson M, De Lima ME, Pimenta AM. Tityus serrulatus venom peptidomics: assessing venom peptide diversity. Toxicon : official journal of the International Society on Toxinology. 2008;52(5):611-8. Epub 2008/08/23.

[74] Shevchenko A, Sunyaev S, Liska A, Bork P, Shevchenko A. Nanoelectrospray tandem mass spectrometry and sequence similarity searching for identification of proteins from organisms with unknown genomes. Methods Mol Biol. 2003;211:221-34. Epub 2002/12/20.

[75] Liska AJ, Shevchenko A. Expanding the organismal scope of proteomics: cross-species protein identification by mass spectrometry and its implications. Proteomics. 2003;3(1):19-28. Epub 2003/01/28.

[76] Sunyaev S, Liska AJ, Golod A, Shevchenko A, Shevchenko A. MultiTag: multiple error-tolerant sequence tag search for the sequence-similarity identification of proteins by mass spectrometry. Analytical chemistry. 2003;75(6):1307-15. Epub 2003/03/28.

[77] Wang L, Evaristo G, Zhou M, Pinkse M, Wang M, Xu Y, et al. Nigrocin-2 peptides from Chinese Odorrana frogs--integration of UPLC/MS/MS with molecular cloning in amphibian skin peptidome analysis. The FEBS journal. 2010;277(6):1519-31. Epub 2010/02/18.

Chapter 3

Computer-Based Methods of Inhibitor Prediction

Silvana Giuliatti

Additional information is available at the end of the chapter

1. Introduction

This chapter presents *in silico* approaches used in protein structure prediction and drug discovery research.

The structural and functional diversity of animal toxins are interesting tools for therapeutic drug design. This diversity is also of great interest in the search for natural or synthetic inhibitors against these animal toxins.

Computational techniques are highly important in drug design. They are used in the search for candidate ligands binding to a receptor.

Drug design based on structure has become a highly developed technology and is used in large pharmaceutical companies. Firstly, the structure of the protein of interest must be known. Therefore, molecular modelling plays an important role in the discovery of new drugs.

If the structure of the receptor is known, then the application is essentially a problem of structure-based drug design. These methods have specific goals, such as attempting to identify the location of the active site of the ligand and the geometry of the ligand in the active site. Another goal is to select a number of related binders in terms of affinity or evaluation of the binding free energy.

The strategy of virtual screening has been used to contribute to the increase in hit rate in the selection of new drug candidates.

Virtual screening (VS) is a modern methodology that has been used in the identification of new bioactive substances. It is an *in silico* method that aims to identify small molecules contained in large databases of compounds with high potential for interaction with target proteins for subsequent biochemical analyses.

The strategy of VS can be divided into *ligand-based virtual screening* (LBVS), where a large number of molecules can be evaluated based on the similarity of known ligands, and *structure-based virtual screening* (SBVS), where a number of molecules can be evaluated for specifically binding to the active sites of target proteins (Figure 1).

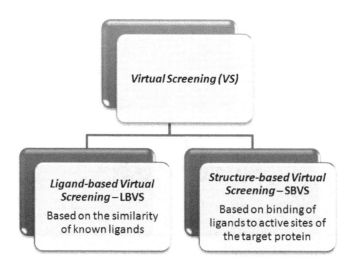

Figure 1. Virtual screening can be divided into ligand-based virtual screening (LBVS) and structure-based virtual screening (SBVS).

Molecular docking is used to determine the best orientation and conformation of a ligand in its receptor site. The aim is to generate a range of conformations of the protein-ligand complex and sort them according to their scores, which are based on their stabilities. In order to do this, the protein structure and a database of ligands (potential candidates) are used as inputs to the docking software. Thus, large collections of virtual compounds are subjected to docking into a protein-binding site and sorted according to their affinities for the macromolecular target, as suggested by the score function.

The focus of this chapter is to present the strategy of SBVS and the basic concepts of the methodologies involved. Examples of these approaches that have been applied to the identification of animal venom inhibitors have been presented at the end of the chapter.

2. Structure-Based Virtual Screening (SBVS)

SBVS involves the evaluation of databases based on the simulation of interactions between the ligands (small molecules) and receptors (target protein). The various steps in the process of SBVS are briefly shown in Figure 2. After obtaining the structure of the receptor and ligand, the next step in the process is molecular docking, which involves the coupling of the ligands with the receptor. At this stage, various conformations and orientations are generated and classified according to the score function. The target protein can be obtained from a database or by modelling.

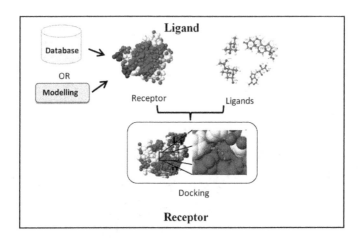

Figure 2. Stages of SBVS. The receptor (the target protein) can be obtained from a database or by modelling. Molecular docking completes the structure-based virtual screening.

2.1. Obtaining the Structure of the Protein Target

Knowledge of the target protein structure is essential for structure-based drug design. The determination of the 3-dimensional structure of the protein may be achieved experimentally by diffraction of X-rays or by magnetic resonance. If the structure of the target protein has already been solved, it can easily be found deposited in public databases such as PDB [37] which contains more than 80,000 experimentally solved structures.

However, sometimes the structure of the target is not known, and this poses a problem in the drug design process. This situation can be resolved by making use of computational methods for predicting protein structure.

Such methods are divided into 2 groups: those based on templates and those that are template-free. The first group includes comparative or homology modelling and threading. The second group includes methods that do not depend on templates to build the model, such as *ab initio* modelling (Figure 3).

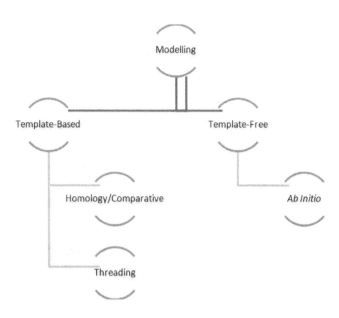

Figure 3. Modelling methods can be classified into template-based methods (homology/comparative modelling) and template-free methods (*ab initio*).

2.1.1. Template-Based Modelling

Homology modelling is based on the use of proteins that share an ancestral relationship with the target protein, that is, that they are evolutionarily related and tend to have similar structures. Thus, this method basically involves knowledge of the primary chain of the target protein and a search among databases for homologous proteins that have solved structures. These proteins are used as templates.

Threading modelling is based on the principle that proteins may have similar structures without sharing the same ancestral relationship because the structure tends to be more conserved than the primary sequence. In this case, these methods evaluate the primary chain of the target protein in relation to proteins that have solved structures.

2.1.1.1. Comparative/Homology Modelling

Comparative or homology modelling constructs a model structure of the target protein using its primary chain and the information obtained from homologous proteins that have solved structures. Therefore, this method depends on the availability of proteins that have structures similar to those of the target and can be used as templates. The whole process requires not only the construction of the model, but also the refinement and evaluation of the obtained model. The process can be divided into stages as follows: selection of the templates, which involves the identification of homologous sequences in a database of proteins that will be used as templates in the modelling process; sequence alignment between the target and the templates; refinement of the alignment; construction of the model, adding loops and side chains; and evaluation of the model (Figure 4).

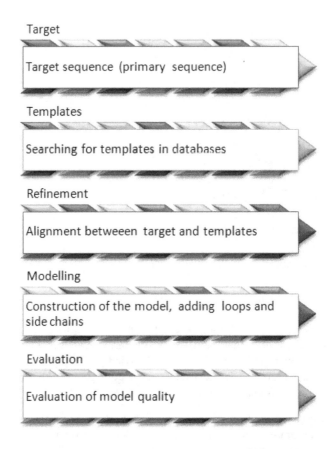

Figure 4. Steps in the comparative modelling process.

The construction of the model depends on the availability of templates. For this purpose, alignment of target and template sequences is widely used and is very efficient. Sequence alignments are typically generated by searching for the result that presents the largest region of identity and similarity. Generally, an identity percentage of at least 25% is considered significant.

There are several tools available for sequence alignment. They differ in the methods used, which can be exhaustive or heuristic, as well as the number of sequences involved in the alignment (multiple or pairwise comparisons). Among these tools, BLAST/PSIBLAST [1; 2] is a tool that performs local alignments based on the profiles between the target sequence and each sequence belonging to a known database.

The results of the alignment can be evaluated using the E-value. The E-value shows an inverse relationship with the identity/similarity between the sequences. Because it is a heuristic method, the results reported by BLAST are generally suboptimal.

If more than 1 template with similar scores is achieved, the best one can be selected as the template with the higher resolution.

Other methods such as HHpred [34] and Pyre [18] use Markov profiles (Hidden Markov models [HMMs]) combined with structural features.

When more than one template is selected, and taking into account that the results are usually suboptimal, there is a need for an alignment between the target protein and the selected templates. In this case, multiple alignments are indicated. There are several tools that perform multiple alignments, such as ClustalW [21]

After obtaining the alignments between the target and templates, the process of obtaining the model of the target protein begins. There are several software tools available, which differ with respect to the method applied. Prominent among these are MODELLER [9, 33] and SWISS-MODEL [3] The software that has shown the best performance is MODELLER. The program models the backbone using a homology-derived restraint method, which is based on the multiple alignment between the target and templates to differentiate between highly conserved and less conserved residues. The model is optimised by energy minimisation and molecular dynamics methods (Figure 5).

Align template structure + target sequence

Extract spatial restraints

Satisfy spatial restraints

Figure 5. The template 3D structures are aligned with the target sequence to be modelled. Spatial features are transferred from the templates to the target and a number of spatial restraints on its structure are obtained. The 3D model is obtained by satisfying all the restraints as thoroughly as possible [33]

The regions of the target that are not aligned with the protein template generally represent loop regions. There are usually some regions caused by insertions and deletions producing gaps in the alignment. Closing these gaps requires modelling of the loops. The loops and the side chains are shaped during the refinement of the model. For this, methods that do not rely on templates can be applied. These include the use of physics parameters and knowledge-based data.

The loops are usually modelled using a database of fragments or by *ab initio* modelling. The use of a database involves finding parts of protein structures known to fit onto 2 regions (stems) of the target protein, which are the regions that precede and follow the loop to be modelled. The conformation of the best matching fragment is used to model the loop.

Ab initio methods generate many random loops and look for one that presents a low-energy state and includes conformational angles contained within the allowed regions of the Ramachandran plot [31] The software CODA [7] can be used for loop modelling.

The side chains can be modelled by programs that make use of libraries of rotamers, such as the software SCRWL4 [20]. The use of rotamer libraries reduces computational time because it reduces the number of favourable torsion angles being examined.

After obtaining the model, its quality must be evaluated. This should be done to make sure that the model has structural features consistent with the physical and chemical rules. Several errors in modelling can occur due to poor choice of template, bad alignment between the target and template, and incorrect determination of loops and side chains.

In the evaluation stage of the model, the structural characteristics as well as the stereochemistry accuracy of the model must be examined.

There are tools available for analysing stereochemical properties, such as PROCHECK [23]. PROCHECK checks the general physicochemical parameters such as phi-psi angles (Ramachandran plot) and chirality. The parameters of the model are compared with those already compiled.

To validate the model for chemical correctness, it is possible to use the software WHAT IF [39]. WHAT IF is a server that checks planarity and bond angles, among other parameters. It also displays the Ramachandran plot.

Verify3D [4, 26] can be used for the analysis of the pseudo-energy profile of the model. It has a database containing environmental profiles based on secondary structures, and the solvent exposure of solved structures at high resolution. It should be noted that the results may be different when different programs are used for verification.

To distinguish correct from incorrect regions, the ERRAT program [6] can be used; this is based on analysis of the characteristics of atomic interactions compared to the highly refined structures.

PROtein Volume Evaluation (PROVE; [30]) calculates the volume of the atoms in the macromolecules using an algorithm that treats the atoms as spheres, analysing the model in relation to the highly resolved and refined structures stored in the PDB.

These software tools are available on servers such as ModFold [27], ProQ (see Section 6 - Table 2), and SAVes (see Section 6 - Table 2).

2.1.1.2. Threading

Threading modelling is generally used when the template and target sequences share less than 30% identity. Thus, structures that do not share an evolutionary relationship with the target protein can be used as templates. However, the target protein has to adopt a fold similar to that of the protein that has had its structure solved. The method can be classified as a pairwise energy-based method.

Using the sequence of the target protein as input, a search is conducted on a database of structures in order to find the best structural match using the criterion of energy calculation. The process is accomplished through a search for solved structures that are most appropriate for the target protein. The comparison highlights secondary structures because they are evolutionarily conserved.

A model is constructed by placing aligned residues between the structure of the template and the target residues. In the next step, the energy of this model is calculated. This is done on various structures in the database. In the end, the models obtained are ranked based on the energy. The model presenting the lowest energy constitutes the most compatible folding model (Figure 6).

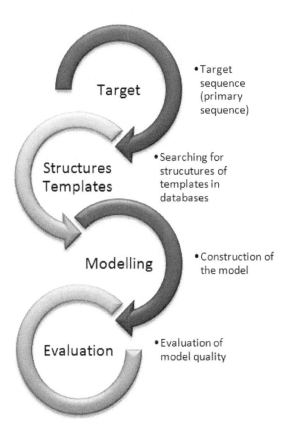

Figure 6. Steps in the threading modelling process.

Many programs such as THREADER [15, 28] and RAPTOR ([41, 42]) can be used to carry out this process.

2.1.2. Template-Free Modelling

One of the biggest problems in comparative modelling is the lack of templates. Template-free methods generate models based on the physicochemical properties and thermodynamic chain of the primary protein target. The processes are iterative. The conformation of the structure is altered until a configuration of lower potential energy is found.

Some methods use force fields based on knowledge as a scoring function. These methods are not strictly free of templates since they employ structures of small fragments of proteins such as, for example, ASTRO-FOLD [19, 35]. Others use energy functions based on first principles of energy and movement of atoms. Generally, these methods involve the calculation of energies of the structures, which has a high computational cost. They are therefore limited to small molecules (approximately 100 residues), as in the case of the software ROSETTA [32].

Firstly, ROSETTA breaks the sequence of the target protein into several short fragments and predicts the secondary structures of the fragments using HMMs. These fragments are then arranged (assembled) into a tertiary setting. Random combinations of these fragments generate a large number of models, which have their energies calculated. The conformation that presents the lowest global energy value is chosen as the best model (Figure 7).

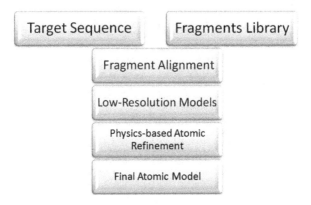

Figure 7. Steps in the ROSETTA process.

3. Molecular Docking

One application of molecular docking is virtual screening, in which a library of compounds is compared to one or more targets, thereby providing an analysis of compounds ranked by potential.

Virtual screening computational techniques are applied to the selection of compounds that can be active in a target protein.

In molecular docking, a ligand is usually placed in the binding site of a predetermined structure of a receptor (Figure 8). In other words, this is a method based on structure. The receptor is typically a protein and the ligand is a small molecule or a peptide. The optimal position and orientation of the ligand are determined using a search algorithm and a scoring function that ranks the solutions.

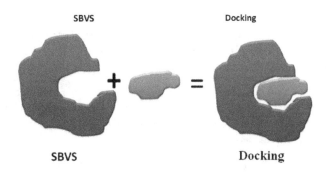

Figure 8. Diagram illustrating the docking of a ligand to a receptor to produce a complex.

The first step of the process of molecular docking is to determine the binding sites of the protein. This can be done by software programs such as Q-Sitefinder [24].

The metaPocket method [13] predicts binding sites using 4 methods: LIGSITEcs [12], PASS [5], Q-Sitefinder, and SURFnet [23] – which in combination increase the success rate of prediction. The methods LIGSITEcs, PASS, and SURFnet use only the geometrical characteristics of the protein structure, detecting regions that have the potential to be binding sites. Such methods do not require prior knowledge of the ligands.

In Q-Sitefinder, the surface of the protein is covered with a layer of methyl probes for the calculation of Van der Waals interactions between the protein and the probe. Probes with favourable interaction energies are retained, and are classified into groups based on the number of probes per group. The largest and most energetically favourable group is ranked first and considered the best potential binding site.

Another step is to define the position of the ligand in the pocket. This can be predicted by molecular docking algorithms.

Several methods have developed different scoring functions and different search methodologies.

The search algorithms have to be able to present different configurations and orientations of the ligand in a short time. Search algorithms, such as those used in molecular dynamics,

Monte Carlo simulations, and genetic algorithms, among others, are all suitable for molecular docking.

Scoring functions must be able to discriminate between different ligand-receptor interactions. These can be grouped into field-force, empirical, and knowledge-based methods.

The algorithms can be classified into rigid body docking and flexible docking algorithms. In rigid-body docking, both the ligand and receptor are rigid. These methods are faster, but do not allow ligand and receptor to adapt to the binding. In flexible methods, the computational cost is higher compared to rigid methods. However, in these cases, the flexibility of the ligand and/or receptor is considered.

Another important factor to be considered in ligand-receptor interactions is the presence of water. Some methods allow water molecules to be positioned. In cases where this is not possible, the position of water molecules can be predicted using a software program such as GRID [17].

GRID calculates the interactions between chemical groups and small molecules with known 3-dimensional structures. The energies are calculated using Lennard-Jones interactions, electrostatic and hydrogen bonding between the compounds, and 3-dimensional structures, using a position-dependent dielectric function.

Examples of tools available for docking proteins include AUTODOCK4.2 [29], GOLD [16], and GLIDE [10].

GOLD uses a genetic algorithm that seeks solutions through docking that propagates multiple copies of flexible models of the ligand in the active site of the receptor and recombining segments of copies at random until a converged set of structures is generated.

The process of searching the databases can be time consuming; a way to reduce the search space is filtering databases by performing a search with the fastest algorithms, selecting the best candidates ranked. Subsequently, within this selection, a search algorithm slowly generates a new ranking of the ligands. Another way to reduce the number of ligands being studied in the database is to perform a search for ligands that offer the greatest possibility of being used in drug design. In this case, it is possible to filter the database by using the AD-MET (absorption, distribution, metabolism, excretion, and toxicity) filter.

Lipinski´s rule of 5 [25] can be used. The rule of 5 is a set of properties that characterise compounds that exhibit good oral bioavailability. It states that, in general, an orally active drug has no more than 1 violation of the rules (Table 1):

Lipinski´s Rule
Not more than 5 hydrogen bond donors (nitrogen or oxygen atoms with one or more hydrogen atoms
Not more than 10 hydrogen bond acceptors (nitrogen or oxygen atoms)
A molecular mass less than 500 daltons
An octanol-water partition coefficient log P not greater than 5

Table 1. Lipinski's Rule of Five

Analysis of the metabolic fate and chemical toxicity of the compounds can be accomplished using the software programs DEREK and METEOR [11]. DEREK predicts whether a given chemical is toxic to humans, mammals, and bacteria. METEOR uses the knowledge of metabolism rules to predict the metabolic fate of chemicals, assisting in the choice of more efficient molecules.

4. Ligand-Based Virtual Screening (LBVS)

Other methods can also be used for screening databases of compounds, such as those based on ligands (LBSV). In this case, a similarity search can be made between known bioactive compounds and molecules contained in databases. LBVS techniques include methods based on the pharmacophore and quantitative structure-activity relationship (QSAR) modelling.

In pharmacophore-based virtual screening, a hypothetical pharmacophore is taken as a template. The goal of screening is to identify molecules that show chemical similarities to the template [40].

QSAR is based on the similarity between structures. It is a quantitative relationship between a biological activity and the molecular descriptors that are used to predict the activity. QSAR searches for similarities between known ligands and each structure in a database, investigating how the biological activity of the ligands can be correlated to their structural features [8].

5. Examples of Virtual Screening / Molecular Docking in Animal Venom

[38] performed a virtual screening against α-Cobratoxin. The neurotoxin α-Cobratoxin (Cbtx), isolated from the venom of the Thai cobra *Naja kaouthia*, causes paralysis by preventing acetylcholine (ACh) binding to nicotinic acetylcholine receptors (nAChRs). A search for α- Cobratoxin structures was carried out in the PDB, and the virtual screening of 1990 compounds was performed using the program AutoDock. On [^{3}H]epibatidine and on [^{125}I] α-bungarotoxin, NSC121865 (compound 23) was most potent in binding with Ac (Kd = 16.26 nM; Kd = 36.63 nM). The results showed that, in clinical applications, NSC121865 would be a very useful potential lead in the development of a new treatment for snakebite victims. This inhibitor can be used for the development of a more potent and specific anti-cobratoxin.

[14] investigated the effects of protease inhibitors, including phenylmethylsulfonyl fluoride (PMSF), benzamidine (BMD), and their derivatives on the activity of recombinant gloshedobin, a snake venom thrombin-like enzyme (SVTLE), from the snake *Gloydius shedaoensis*. The structural model of gloshedobin was built by homology modelling using modelling package MODELLER. The stereochemical quality of the homology model was assessed using the PROCHECK program and the software AutoDock was used to dock inhibitors onto the structural model of gloshedobin. The docking results indicated that the strongest inhibitor, PMSF, bound covalently to the catalytic Ser195.

[36] evaluated the inhibitory effect of 1-(3-dimethylaminopropyl)-1-(4-fluorophenyl)-3-oxo-1,3-dihydroisobenzofuran-5-carbonitrile (DFD) on viper venom-induced haemorrhagic and PLA2 activities. Molecular docking studies of DFD and snake venom metalloproteases (SVMPs) were performed to understand the mechanism of inhibition by DFD, since SVMPs constitute one of the protein groups responsible for venom-induced haemorrhage. The docking results showed that DFD binds to a hydrophobic pocket in SVMPs with the Ki of 19.26 x 10 $^{-9}$ (kcal/mol) without chelating Zn2+ in the active site.

6. Conclusions

In silico approaches used in protein structure prediction and in drug discovery research have been presented in this chapter.

Computational methods used in the search for inhibitors play an essential role in the process of discovering new drugs.

The application of protein modelling methods has contributed significantly in cases where the structure of the target protein has not been solved, allowing the SBVS process be completed.

Good results obtained by virtual screening depend on the quality of structures, databases to be scanned, the search algorithms, and scoring functions. Therefore, there must be a good interaction and exchange of information between *in silico* and experimental methods. Careful application of these strategies is necessary for successful drug design.

Table 2 presents a list of software tools and server web sites.

Summary Tools	
PDB	http://www.rcsb.org/pdb/home/home.do
BLAST	http://blast.ncbi.nlm.nih.gov/
HHpred	http://toolkit.tuebingen.mpg.de/hhpred
ClustalW	http://www.ebi.ac.uk/Tools/msa/clustalw2/
SWISS-MODEL	http://swissmodel.expasy.org/
MODELLER	http://salilab.org/modeller/
SCRWL4	http://dunbrack.fccc.edu/scwrl4/
PROCHECK	http://www.ebi.ac.uk/thornton-srv/software/PROCHECK/
WHAT IF	http://swift.cmbi.ru.nl/whatif/
Verify3D	http://nihserver.mbi.ucla.edu/Verify_3D/
ERRAT	http://nihserver.mbi.ucla.edu/ERRATv2/
PROVE	http://www.doe-mbi.ucla.edu/Software/PROVE.html

Summary Tools	
modFold	https://www.reading.ac.uk/bioinf/ModFOLD/
ProQ	http://www.sbc.su.se/~bjornw/ProQ/ProQ.html
ROSETTA	http://www.rosettacommons.org/home
Q-sitefinder	http://www.modelling.leeds.ac.uk/qsitefinder/
SAVes	http://nihserver.mbi.ucla.edu/SAVES/
THREADER	http://bioinf.cs.ucl.ac.uk/software_downloads/threader/
metaPocket	http://projects.biotec.tu-dresden.de/metapocket/
PASS	http://www.ccl.net/cca/software/UNIX/pass/overview.shtml
SURFNET	http://www.ebi.ac.uk/thornton-srv/software/SURFNET/
AUTODOCK	http://autodock.scripps.edu/
GOLD	http://www.ccdc.cam.ac.uk/products/life_sciences/gold/
GLIDE	http://www.schrodinger.com/products/14/5/
Derek/Meteor	https://www.lhasalimited.org/
Raptorx	http://raptorx.uchicago.edu/
RAPTOR	http://www.bioinformaticssolutions.com/raptor/downloadpricing/freetrial.html
Phyre	http://www.sbg.bio.ic.ac.uk/~phyre/
MUSTER	http://zhanglab.ccmb.med.umich.edu/MUSTER/
I-TASSER	http://zhanglab.ccmb.med.umich.edu/I-TASSER/

Table 2. Software tools and server web sites.

Acknowledgements

The author would like to thank CAPES-PROEX and CNPq for financial support.

Author details

Silvana Giuliatti*

Address all correspondence to: silvana@fmrp.usp

Faculty of Medicine of Ribeirão Preto - University of São Paulo, Brazil

References

[1] Altschul, S. F., Madden, T. L., Schäffer, A. A., Zhang, J., Zhang, Z., Miller, W., & Lipman, D. (1997). Gapped BLAST and PSI-BLAST: A New Generation of Protein Database Search Programs. *Nucleic Acids Research*, 25(17), September, 3389-3402, 1362-4962.

[2] Altschul, S. F., Gish, W., Miller, W., Myers, E. W., & Lipman, D. J. (1990). Basic Local Alignment Search Tool. *Journal of Molecular Biology*, 215(3), October, 403-410, 0022-2836.

[3] Arnold, K., Bordoli, L., Kopp, J., & Schwede, T. (2006). The SWISS-MODEL Workspace: a Web-Based Environment for Protein Structure Homology Modelling. *Bioinformatics*, 22(2), January 2005, 195-201, 1460-2059.

[4] Bowie, J. U., Lüthy, R., & Eisemberg, D. (1991). A Method to Identify Protein Sequences that Fold into a Known Three-Dimensional Structure. *Science*, 253(5016), July, 164-170, 0036-8075.

[5] Brady, G., & Stouten, P. (2000). Fast Prediction and Visualization of Protein Binding Pockets with PASS. *Journal of Computer-Aided Molecular Design*, 14(4), May, 383-401, 1573-4951.

[6] Colovos, C., & Yeates, T. O. (1993). Verification of Protein Structures: Patterns of Nonbonded Atomic Interactions. *Protein Science*, 12(9), September, 1511-1519, 0036-8075.

[7] Deane, C. M., & Blundell, T. L. (2001). CODA: A Combined Algorithm for Predicting the Structurally Variable Regions of Protein Models. *Protein Science*, 10(3), March, 599-612, 0146-9896 X.

[8] Ebalunode, J. O., Zheng, W., & Tropsha, A. (2011). Application of QSAR and Shape Pharmacophore Modeling Approaches for Targeted Chemical Library Design. *Methods in Molecular Biology*, 685, 111-133, 1064-3745.

[9] Eswar, N., Marti-Renom, M. A., Webb, B., Madhusudhan, M. S., Eramian, D., Shen, M., Pieper, U., & Sali, A. (2007). Comparative Protein Structure Modelling With MODELLER. *Current Protocols in Bioinformatics*, 50, (November), unit 2.9.1-2.9.31, 1934-340X.

[10] Friesner, R. A., Banks, J. L., Murphy, R. B., Halgren, T. A., Klicic, J. J., Mainz, D. T., Repasky, M. P., Knoll, E. H., Shaw, D. E., Shelley, M., Perry, J. K., Francis, P., & Shenkin, P. S. (2004). Glide: A New Approach for Rapid, Accurate Docking and Scoring. 1. Method and Assessment of Docking Accuracy. *Journal of Medical Chemistry*, 47(7), March, 1739-1749, 1520-4804.

[11] Greene, N., Judson, P., Langowski, J., & Marchant, C. A. (1999). Knowledge-based expert Systems for Toxicity and Metabolism Prediction: DEREK, StAR and METEOR. *SAR QSAR Environmental Research*, 10(2-3), 299-313, 0013-9351.

[12] Huang, B., & Schroeder, M. (2006). LIGSITEcsc: Predictiong Ligand Binding Sites using the Connolly Surface and Degree of Conservation. *BMC Structural Biology*, 6, September, 19, 1472-6807.

[13] Huang, B. (2009). MetaPocket: A Meta Approach to Improve Protein Ligand Binding Site Prediction. *OMICS: A Journal of Integrative Biology*, 13(4), August, 325-330, 1557-8100.

[14] Jiang, X., Chena, L., Xua, J., & Yanga, Q. (2010). Molecular Mechanism Analysis of Gloydius Shedaoensis Venom Gloshedobin. *International Journal of Biological Macromolecules*, 48(1), January, 129-133, 0141-8130.

[15] Jones, D. T., Taylor, W. R., & Thornton, J. M. (1992). A New approach to Protein Fold Recognition. *Nature*, July, 358, 86-96, 0028-0836.

[16] Jones, G., Willett, P., Glen, R. C., Leach, A. R., & Taylor, R. (1997). Development and Validation of a Genetic Algorithm for Flexible Docking. *Journal of Molecular Biology*, 267(6381), July, 727-748, 0022-2836.

[17] Kastenholz, M. A., Pastor, M., Cruciani, G., Haaksma, E. E. J., & Fox, T. (2000). GRID/CPCA: A New Computational Tool to Design Selective Ligands. *Journal of Medical Chemistry*, 43(16), August, 3033-3044, 1520-4804.

[18] Kelley, L. A., & Stemberg, J. E. (2009). Protein Structure Prediction on the Web: a Case Study using the Phyre Server. *Nature Protocols*, 4(3), February, 363-371, 1754-2189.

[19] Klepeis, J. L., & Floudas, C. A. (2003). ASTRO-FOLD: A Combinatorial and Global Optimization Framework for Ab Initio prediction of Three-Dimensional Structures of Proteins from the Amino Acid Sequence. *Biophysical Journal*, 85(4), October, 2119-2146, 0006-3495.

[20] Krivov, G. G., Shapovalov, M. V., & Dunbrack, R. L. (2009). Improved Prediction of Protein Side-Chain Conformations with SCWRL4. *Proteins*, 77(4), December, 778-795, 1097-0134.

[21] Larkin, M. A., Blackshields, G., Brown, N. P., Chenna, R., Mc Gettigan, P. A., Mc William, H., Valentin, F., Wallace, I. M., Wilm, A., Lopez, R., Thompson, J. D., Gibson, T. J., & Higgins, D. G. (2007). Clustal W and Clustal X Version 2.0. *Bioinformatics*, 23(21), November, 2947-2948, 1460-2059.

[22] Laskowiski, R. (1995). SURFNET: a Program for Visualizing Molecular Surfaces, Cavities and Intermolecular Interactions. *Journal of Molecular Graphics*, 13(5), October, 323-330, 0263-7855.

[23] Laskowski, R. A., Macarthur, M. W., Moss, D. S., & Thornton, J. M. (1993). PROCHECK: a Program to Check the Stereochemical Quality of Protein Structures. *Journal of Applied Crystallography*, 26(2), April, 283-291, 1600-5767.

[24] Laurie, A., & Jackson, R. (2005). Q-SiteFinder: an Energy-based Method for the Prediction of Protein-Ligand Binding Sites. *Bioinformatics*, 21(9), May, 1908-1916, 1046-2059.

[25] Lipinski, C. A., Lombardo, F., Dominy, B. W., & Feeney, P. J. (2001). Experimental and Computational Approaches to Estimate Solubility and Permeability in Drug Discovery and Development Settings. *Advanced Drug Delivery Reviews*, 46(1-3), March, 3-26, 0016-9409 X.

[26] Lüthy, R., Bowie, J. U., & Eisemberg, D. (1992). Assessment of Protein Models with Three-Dimensional Profiles. *Nature*, 356(6364), March, 83-85, 0028-0836.

[27] Mc Guffin, L. J. (2008). The ModFOLD Server for the Quality Assessment of Protein Structural Models. *Bioinformatics*, 24, 586-587, 1460-2059.

[28] Milleer, R. T., Jones, D. T., & Thornton, J. M. (1996). Protein Fold Recognition by Sequence Threading: Tools and Assessment Techniques. *The FASEB Journal*, 10(1), January, 171-178, 1530-6860.

[29] Morris, G. M., Huey, R., Lindstrom, W., Sanner, M. F., Belew, R. K., Goodsell, D. S., & Olson, A. J. (2004). AutoDock4 and AutoDockTools4: Automated Docking with Selective Receptor Flexibility. *Journal of Computational Chemistry*, 30(16), December,2009, 2785-2791, 0109-6987 X.

[30] Pontius, J., Richelle, J., & Wodak, S. J. (1996). Deviations from Standard Atomic Volumes as a Quality Measure of Protein Crystal Structures. *Journal of Molecular Biology*, 264(1), November, 121-126, 0022-2836.

[31] Ramachandran, G. N., Ramakrishnan, C., & Sasisekharan, V. (1963). Stereochemistry of Polypeptide Chain Configurations. *Journal of Molecular Biology*, 7, July, 95-99, 0022-2836.

[32] Rohl, C. A., Strauss, C. E., Misura, K. M. S., & Baker, D. (2004). Protein Sructure Prediction using Rosetta. *Methods Enzymol*, 383, 66-93, 0076-6879.

[33] Sali, A. E., & Blundell, T. L. (1993). Comparative Protein Modelling by Satisfaction of Spatial Restraints. *Journal of Molecular Biology*, 234, 779-815, 0022-2836.

[34] Söding, J., Biegert, A., & Lupas, A. N. (2005). The HHpred Interactive Server for Protein Homology Detection and Structure Prediction. *Nucleic Acids Research*, 33(3), December, W244-W248, 1362-4962.

[35] Subramani, A., Wei, Y., & Floudas, C. A. (2012). ASTRO-FOLD 2.0: An Enhanced Framework for Protein Structure Prediction. *American Institute of Chemical Engineers Journal*, 58(5), May, 1619-1637, 1547-5905.

[36] Sunitha, K., Hemshekhar, M., Gaonkar, S. L., Santhosh, M. S., Kumar, M. S., Basappa Priya, B. S., Kemparaju, K., Rangappa, K. S., Swamy, S. N., & Girish, K. S. (2011). Neutralization of Hanemorrhagic Activity of Viper Venoms by 1-(3-Dimethylamino-

propyl)-1-(4-Fluorophenyl)-3-Oxo-1, 3-Dihydroisobenzofuran-5-Carbonitrile. *Basic & Clinical Pharmacology & Toxicology*, 109(4), October, 292-299, 1742-7843.

[37] Sussman, J. L., Lin, D., Jiang, J., Manning, N. O., Prilusky, J., Ritter, O., & Abola, E. E. (1998). Protein data bank (PDB): a Database of 3D Structural Information of Biological Macromolecules. *Acta Crystal*, D54, 1078-1084, 1600-5759.

[38] Utsintong, M., Talley, T. T., Taylor, P. W., Olson, A. J., & Vajragupta, O. (2009). Virtual Screening Against α-Cobratoxin. *Journal of Biomolecular Screening*, 14(9), October, 1109-1118, 1087-0571.

[39] Vriend, G. (1990). WHAT IF: A Molecular Modelling and Drug Design Program. *Journal of Molecular Graphics*, 8(1), March, 52-56, 0263-7855.

[40] Yang, U. S. (2010). Pharmacophore Modeling and Applications in Drug Discovery: Challenges and Recent Advances. *Drug Discovery Today*, 15(11-12), June, 446-450, 1359-6446.

[41] Peng , J., & Xu, J. (2010). Low-homology protein threading. *Bioinformatics*, 26, i294-i300, 10-1093.

[42] Peng , J., & Xu, J. (2011). RaptorX: Exploiting Structure Information for protein alignment by statistical inferenc. *Proteins: Structure, Functon, and Bioinformatics*, 79(S10), 167-171, 10-1002.

Chapter 4

New Perspectives in Drug Discovery Using Neuroactive Molecules From the Venom of Arthropods

Márcia Renata Mortari and
Alexandra Olimpio Siqueira Cunha

Additional information is available at the end of the chapter

1. Introduction

Arthropods are one of the most ancient groups of animals in earth and their venoms have been responsible for their chemical defense in a very efficient way. Resulting from an intense and elaborated evolutionary process, venoms produced by arthropods have a very complex repertoire of biologically active molecules. When inoculated in mammals these molecules induce a wide range of systemic effects, including actions in the CNS. In mammalian CNS, venom compounds may either inhibit or stimulate with affinity and specificity structures such as: ion channels, neurotransmitter receptors and transporters [1-3]. Not surprisingly, these actions have attracted the attention of many investigators in search of tools to help the understanding of neural mechanisms as well as those in search of novel probes in CNS drug design for the last 20 years [3,4]. In addition to the growing interest in finding new neuroactive compounds, the improvement of proteomic and transcriptome techniques has stimulated great progress in the bioprospecting, enabling and accelerating the testing of new toxins in several animal models. Animal research aiming at the efficacy of peptides and acylpoliamines, isolated from arthropod venoms, have revealed the great potential of these compounds to treat various diseases, such as epilepsy, Parkinson's, Alzheimer's, chronic pain and anxiety disorders

According to World Health Organization (WHO), neurological and mental disorders are one of the greatest threats to public health not only for its direct and immediate effects, but also for the progressive nature of these diseases, often leading to disability and death [5]. The symptoms of most of these diseases are often well treated with a several pharmaceuticals, such as antidepressants, anxiolytics, anticonvulsants and analgesics. However, it is well known that neuroactive drugs may induce a complex range of adverse effects that limit the usage in some patients or may even function as a factor of impairment in people's quality of life. According to [6], none of antiepileptic drugs discovered in the last 20 years, was efficient to cure or even suppress seizures in epileptic patients. Therefore, there is a continued need for the discovery of novel drugs to treat most neurological and mental disorders [7].

This chapter will target the discussion of recent contributions of research on the compounds of arthropod venom, for the discovery of novel tools to study the functioning of the structures of mammalian CNS, as well as the supply of novel alternatives to the treatment of neurological disorders. Among the major compounds, it will be highlighted those with the analgesic, anxiolytic, antiepileptic and neuroprotective effects, with emphasis on the most promising on preclinical or clinic evaluation.

2. Main targets of the neuroactive compounds isolated from arthropod venoms

Venom isolated from bee, scorpion and spider have been used to the treatment of various diseases in Chinese and Korean traditional medicine, such as epilepsy, stroke, facial paralysis, arthritis, rheumatism, back pain, cancerous, tumors, and skin diseases [8-10]. Moreover, venoms of arthropod animals have been used to study various physiopathological processes, and also offer opportunity to design and develop new therapeutic drugs [3,11,12] .

Arthropod venoms are rich in biologically active substances with different physiological actions, specially the neurotoxins. So far, identified neurotoxins generally comprise the classes of peptides or acylpolyamines, acting with affinity and specificity over excitatory or inhibitory neurotransmissions (for revision see [12]. The actions of these compounds include the interaction with Na^+, K^+ and Ca^{2+} ion channels, agonism or antagonism of metabotropic and ionotropic receptors for neurotransmitters as the excitatory neurotransmitter glutamate. At the presynaptic level, several studies have shown the interaction of arthropod neurotoxins with protein transporters of neurotransmitters, resulting in the facilitation or inhibition of their uptake.

3. Antinociceptive effects

Of extreme importance for the organism, pain is an indicator of corporal integrity and has been considered since January 2000, by the Joint Commission on Accreditation on Healthcare Organizations (JCAHO) as the fifth vital sign that should be assessed and recorded together with other signals immediately after birth. According to the International Association for the Study of Pain (IASP), pain is defined as an unpleasant sensation and emotional experience associated with actual or potential tissue damage. However, approximately one third of world population suffers from pathological persistent or recurrent pain, which is a common complaint in patients with different diseases, and exerts great impact on their social life [13]. In these cases, treatment is a challenge for researchers and health professionals who constantly seek new therapeutic strategies, since most of these are inadequate or cause serious side effects [14].

Analgesics and systemic conservative therapies are widely used for pain control. However, in many cases, especially in patients with neuropathic pain, more aggressive treatments are needed, which promote a significant clinical improvement but only in 30-50% of patients [15,16].

Although an injection of arthropod venoms is commonly reported to cause tonic pain and hyperalgesia, there is also evidence suggesting that these venoms might have antinociceptive effects on inflammation. Thus, nowadays, toxins isolated from arthropods are considered powerful tools, since they have congruent targets of the impulse transmission of pain, and may provide an attractive alternative to opioid treatments.

3.1. Polypeptide toxins from Scorpion

The most studied Arthropod venom is extracted from the Asian scorpion *Mesobuthus martensi* Karsch (BmK). It is composed of several toxins, and so far, ten have been described, which produce powerful antinociceptive effects. This is the case of the two β-excitatory anti-insect toxins BmK IT-AP (or Bm33-I) and BmK AngP1, two β- depressant anti-insect toxins BmK dITAP3 and BmK IT2, as well as six toxins yet without consensus classification, BmK AS, BmK AS1, BmK AGAP, BmK Ang M1, BmK AGP-SYPU1 and BmK AGP-SYPU2. These compounds probably belong to a family of peptides NaScTx that are composed of 60-76 amino acid residues with four disulfide bonds, the cysteine positions among these toxins are highly conserved [17,18]. Considering their structures, they might be able to bind to sodium channels impairing depolarization of the action potential in nerve and muscle, resulting in neurotoxicity [18], although it remains to be fully investigated.

The NaScTx family can be classified in at least two major families, α and β, according to the mode of action on Na^+ channels [19]. The binding of α-toxins delays Na_v channel inactivation, while that of β-toxins shifts the membrane potential dependence of channel activation to more negative potentials. α and β-toxins also exhibit pharmacological preferences for mammals or insects sodium channels. Therefore, considering their pharmacological activities, α and β NAScTx can be also divided into three groups:

i.   "classic" highly specific for mammals;

ii.   "α-like toxins" active both on mammals and insects, which are far less specific and less active than the "classical" ones;

iii. α-toxins only specific for insects and without any toxicity on mammals, even at high concentrations. Moreover, the insect selective β-toxins have been divided into two groups: the excitatory insect toxins and the depressant insect toxins.

Regarding the β-excitatory anti-insect toxins, BmK IT-AP (Insect Toxin-Analgesic Peptide), which was isolated in 1999, produces a potent antinociceptive effect in mouse-twisting model, after i.v. injection [20]. The same toxin has also been sequenced by another group and named Bm K 33-I [21]. Later, Guan and colleagues [22] identified a novel toxin with analgesic effects, BmK AngP$_1$, which shows an evident analgesic effect with simultaneous excitatory insect toxicity, but is devoid of any toxicity on mice even at high dosages. The analgesic effect was assessed with a mouse-twisting model. The analgesic effect on mice of the $AngP_1$ is at least 4-5 times weaker than that of IT-AP, but the toxicity to insects is twice as strong as that of IT-AP [20,22]

In relation of depressant toxins isolated from BmK venom, BmK IT2 has been more studied from the venom of BmK (Fig 1). Intraplantar injection of BmK IT2 inhibited thermal hyperalgesia in carrageenan-treated rats and significantly prolonged paw withdrawal latency in normal rats [23]. This toxin also displays an inhibitory effect on the C component of the rat nociceptive flexion reflex by subcutaneous injection in vivo [24]. Peripheral or spinal delivery of BmK IT2 suppressed formalin-induced nociceptive behaviors and c-Fos expression in spinal cord [25,26]. Both BmK IT2 and Bm K dIT-AP3 (depressant Insect Toxin-Analgesic Peptide 3) are toxic for insects, but not for mammals [27], and shows 86.7% of sequence similarity [23]. BmK dIT-AP3 also induces analgesia in the mouse-twisting model [18]. Using whole-cell patch clamp, it has been shown that BmK dIT-AP3 inhibits Na_v currents of rat dorsal root ganglion (DRG) neurons, blocking more selectively the tetrodotoxin-resistant (TTX-R) component of the Na^+ currents. These results suggest that the inhibition of the rat nociceptive flexion reflex by BmK dITAP3 may be attributed to modulation of the DRG's voltage-gated Na^+ channels [24].

Wang and colleagues [28] isolated a new antinociceptive peptide, named BmK AGP-SYPU1. Recombinant BmK AGP-SYPU1 showed similar analgesic effects on mice compared to natural when assayed using a mouse-twisting model [28]. More recently, BmK AGP-SYPU2 was purified and tested, also in mouse-twisting model. Sequence determination showed that the mature BmK AGP-SYPU2 peptide is composed of 66 amino acid residues, and BmK AGP-SYPU2 is identical to BmK alpha2 and BmK alphaTX11.

BmK AS had a strong analgesic effect on both visceral and somatic pain [29,30]. It relieves formalin-induced two-phase spontaneous flinching response and carrageenan-induced mechanical hyperalgesia, probably by modulating the voltage-gated Na^+ channels of sensory neurons [31,32]. Moreover, BmK AS showed activity nearly equivalent to that of morphine. Later, a new peptide that possesses 86.3% of similarity with BmK AS was identified. Both polypeptides have 66 amino acids cross-linked by four disulfide bridges [29]. In addition, these two peptides show a poor similarity with other known types of scorpion toxins. BmK AS and AS1 are not toxic against mammals and only have a weak toxicity to insects. BmK AS, then BmK AS1, have been found to significantly stimulate the binding of [3H]-ryanodine to partially purified ryanodine receptors [33]. More recently, electrophysiological studies have shown that they are able to inhibit Na^+ currents in NG108-15 cells [34] and to depress TTX-sensitive and TTX-resistant Na^+ currents in rat small DRG neurons. Interestingly, in rat models, BmK AS1 also displays antinociceptive effects according to [33]. These authors concluded that the effects could be mediated by the modulation of voltage-gated Na^+ channels and they also suggested that BmK AS and BmK AS1 could form a new family of scorpion insect toxins.

BmK AGAP (antitumor-analgesic peptide), isolated in 2003, had strong inhibitory effect on both viscera and soma pain [35]. To evaluate the extent to which residues of the toxin core contribute to its analgesic activity, nine mutants of BmK AGAP were produced and tested. However, further studies are necessarily to elucidate the mechanism of action as well as to exploit its analgesic activity [36]. In relation to BmK Ang M1 [37], it also was reported to exhibit potential analgesic effect. Moreover, electrophysiological studies showed that BmK AngM1 at the concentration of 1 µM inhibited voltage-dependent Na^+ current (I_{Na}) and voltage-dependent delayed rectifier K^+ current (I_K), but had no effects on transient K^+ current [37].

It is important to note that the excitatory and depressant anti-insect toxins belong to different groups, which have distinct modes of interaction with receptors. Thus, one can infer that the analgesic effect of these peptides may have a molecular mode and mechanism different from that of insect toxicity. Still, the mechanisms by which these scorpion toxins can modulate pain pathways remain to be clarified. According to [8], four different possibilities might be described:

i.   peptides act directly Na^+ channels involved in the pathway of pain,

ii.   peptides modulate indirectly the pain sensation,

iii.   peptides also modulate other targets involved in pain pathway

iv.   pain alleviation is only apparent and results from misinterpretations that might have occurred from animal models used.

3.2. Polypeptide toxins from Spider

Another group of arthropods that have very promising antinociceptive compounds are spiders [41]. In 1996, Roerig & Howse reported the effect of ω-agatoxina IVA (Fig 1) isolated from funnel spider *Agelenopsis aperta* venom, against thermal stimulation in the tail flick test, when co-administrated with morphine intrathecal. Intrathecal injection of ω-agatoxin IVA (0.2 nmol/kg) also decreased the licking time in both the early and late response phases in a dose-dependent manner in the Formalin test [42]. The use of this peptide as an analgesic could be of particular benefit in patients tolerant or opioid-dependent, since this compound exhibits selectivity for the P/Q Ca^{2+} channels [43]. Other spider venom very promissory is the venom of the Brazilian armed spider *Phoneutria nigriventer*, the purified fraction 3 (PhTx3) contains 6 toxin isoforms (Tx3-1 to -6) [44,45] that target Ca^{2+} channels with different affinity patterns. Moreover, one toxin, Tx3-6 (Phα1β), demonstrated that it preferentially blocks the N-type calcium current [46] and produce a potent antinociceptive effect with higher therapeutic index [44]. Dalmolin and colleagues [45] showed that Tx3-3 (purified the same fraction) caused a short-lasting antinociceptive effect in the nociceptive pain test and a long-lasting antinociceptive effect in neuropathic pain models, without producing detectable side effects. However, Tx3-3 did not change the inflammatory pain. Tx3-3 blockade of P/Q- and R-type Ca^{2+} channels and inhibit the glutamate release in rat brain cortical synaptosomes [47]. Other neurotoxin isolated from spider *Phoneutria nigriventer* is Phα1β, which is a potent toxin blocking neuronal voltage-sensitive Ca^{2+} channels. This peptide induced longer antiallodynic effect than µ-conotoxin MVIIA and morphine in mice [48].

In addition to toxins calcium modulators, compounds isolated from spider that interact with other ionic channels have shown great potential. A new class of peptide toxins named is the Huwentoxin I (HWTX-I, Fig 1) that is the most abundant toxic component in the crude venom of the Chinese bird spider *Ornithoctonus huwena*. Whole-cell patch clamp records revealed that HWTX-I selectively inhibits N-type Ca^{2+} channels in NG108-15 cells, and it also can block transmitter release from nerve endings by preventing depolarization induced by calcium influx [38] Antinociception effect of the HWTX-I in formalin test was greater and lasted two-fold longer time compared to morphine [39]. Furthermore, Tao and collaborators [40] demonstrated that intrathecal administration of HWTX-I is effective in antinociception in the rat model of rheumatoid arthritis more effective than ibuprofen.

Several studies have reported that intrathecal administration of non-selective blockers of Ca^{2+} channels shows antinociceptive effects in animals tested with thermal stimuli: hot plate and tail flick. According to [49], N and P/Q Ca^{2+} channels are probably involved in nociceptive behavior induced by formalin injection in rats, while the L-type channels has no effect. N- and P/Q-type Ca^{2+} channels are expressed specifically in the nervous system, and they have a major importance in controlling the excitation of spinal neurons from sensory afferents of inflamed tissues, relieving inflammatory pain.

A new class of peptide toxins named π-theraphotoxin-Pc1a (π-TRTX-Pc1a; also known as psalmotoxin-1 (PcTx1) was isolated from the venom of the spider neotropical *Psalmopoeus cambridgei* (Fig.1). π-TRTX-Pc1a is the most potent and selective blocker of ion channels sensitive to acid – ASICa [50]. These channels play important roles in pathological conditions such as cerebral ischemia or epilepsy, as well as being responsible for the sensation of pain that accompanies tissue acidosis and inflammation [51]. Since external acidification is a major factor in pain associated with inflammation (hematosis muscle and cardiac ischemia, or cancer), these neurotoxins can be used to control the pain sensation triggered by these channels [52]. π-TRTX-Pc1a was shown to be an effective analgesic, comparable to morphine, in rat models of acute and neuropathic pain when injected directly in Central Nervous System [53] and intranasal administration of this peptide resulted in neuroprotection of neurons in a mouse model of ischemic stroke even when administered hours after injury [54].

Other important target in the search for new analgesics isolated from spider venoms are Na_V channels, since modulatory compounds of these channels are the dominant pharmacological species in spider venoms, although still poorly characterized. In this context, Intrathecal administration of β-TRTX-Gr1b (formerly GsAFI), a peptide obtained from venom of *Grammostola spatulata*, the Chilean pink tarantula spider, induced analgesia in a variety of rat pain models such as the tail flick latency test, hot plate threshold test, von Frey threshold test, and formalin pain test, without any confounding side-effects. Moreover, the β-TRTX-Gr1b peptide did not exhibit cross tolerance with morphine [55].

Further on spider venoms, Purotoxin-1 (PT1) was recently isolated the, from the venom of the Central Asian spider *Geolycosa sp* [56]. PT1 is a 35-residue peptide with four disulfide bonds, and it exerts a potent analgesic effect in rat models of acute and chronic inflammatory pain by injection of either carrageenan or Freund's complete adjuvant, respectively. PT1 was also effective in reducing the number of nocifensive events triggered by the injection of capsaicin or formalin (only second phase) [56]. This molecule also inhibits P2X3 receptors in a powerful and selective manner. These ATP-activated receptors are largely expressed in mammalian sensory neurons play a key role in the pain perception. Thus, PT1 appears to be a promising lead compound for the development of analgesics that target these receptors [56].

3.3. Polypeptide toxins from Bees and Wasps

Bee venom has been traditionally used to relieve pain and treat chronic pain diseases (for revision see [57]). Moreover, acupoint stimulation into the subcutaneous region (acupuncture) rather than other injection sites may be important for the antinociceptive effects of this venom. There is increasing evidence suggesting that bee venom has antinociceptive effects on visceral nociceptive effects, mechanical and thermal hyperplasia, formalin-induced pain behavior and collagen-induced arthritic pain, as well as knee osteoarthritis (OA)-related pain [58-63]. BV contains at least 18 active components, including enzymes, peptides, and biogenic amines, which have a wide variety of pharmaceutical properties, and so multiple mechanisms associated to antinociceptive effects have been suggested, such as activation of the central and spinal opioid receptor and α_2-adrenergic receptor, as well as activation of the descending serotonergic pathways (for revision see [64]).

Melittin is a small protein containing 26 amino acid residues and is the major bioactive component in BV (Fig.1). This polypeptide readily integrates into and disrupts both natural and synthetic phospholipid bilayers [65,66]. Melittin also enhances the activity of PLA_2 [67] and has a variety of effects on living cells possibly through the disruption of the membrane [68]. The decrease in cyclooxygenase (COX)-2 and phospholipase PLA_2 expression and the decrease in the levels of tumor necrosis factor alpha (TNF-α), interleukin (IL)-1, IL-6, nitric oxide (NO) and oxygen reactive species (ROS) are suggested to be associated with the anti-arthritis effect of melittin [69]. This peptide has also been thought to play a role in production of anti-nociceptive and anti-inflammatory effects [64]. In addition, Merlo and colleagues [70] demonstrated the antinociceptive activity of the melittin in experimental models of nociceptive and inflammatory pain. Interestingly, melittin failed to increase the latency for the nociceptive response in the hot-plate model and in the first phase of the formalin test, revealing that melittin presents an activity that resembles more that of anti-inflammatory drugs and less that of centrally acting drugs [70]. Nevertheless, the molecular and cellular mechanisms underlying the anti-nociceptive effects of melittin are not entirely clear and remain to be further clarified by further experimental studies [57].

Addition of melittin, adolapin has been isolated from BV and it demonstrated a potent analgesic effect in mouse-twisting model and the Randall-Sellito's test [71]. The anti-inflammatory activity of adolapin was evaluated and it had a pronounced activity in the following tests: carrageenan, PG, adjuvant rat hind paw edema and adjuvant polyarthritis. The effects of adolapin are presumably due to its ability to inhibit the prostaglandin synthesis via inhibition of cyclooxygenase activity [71,72].

Venoms of wasps also have analgesic peptides. Mortari and colleagues [73] isolated a compound with antinociceptive activity from the venom of the Brazilian social wasp *Polybia occidentalis*. The isolated peptide is a neurokinin named Thr[6]-Bradykinin. This neurokinin is a small peptide consisting of nine amino acid residues, Arg-Pro-Pro-Gly-Phe-Thr-Pro-Phe-Arg-OH, which exhibits a high degree of homology with bradykinin (BK), except for the substitution of Thr for Ser in position 6 at BK. As a result, small changes in their secondary structures are observed [74]. This modification has been regarded as responsible for increasing B_2 receptor affinity and potency of Thr[6]-BK in relation to BK *in vitro* and *in vivo* [74,75]. Thr[6]-BK antinociceptive effect was dose- and time- dependent, when injected directly into the CNS of rats in hot-plate and tail-flick tests, and it was three times more potent than morphine and 4 times more potent than BK in tail-flick test. Thr[6]-BK induced antinociception by activating presynaptic B_2 receptors, which activate descending adrenergic pathways. Studies investigating the role of kinins in the CNS provide new information on the supraspinal system of the pain control, whose modulation may represent a new strategy to control pain-related pathologies [76].

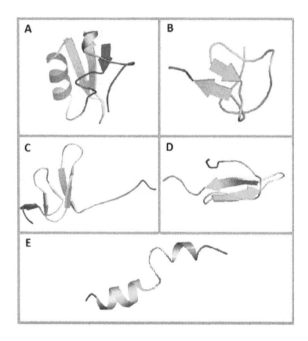

Figure 1. Tridimensional structure of antinociceptive peptides isolated from arthropod venoms. (A) BMK IT2; (B) HWTX 1; (C) ω-Agatoxin IVA; (D) π-Theraphotoxin-Pc1a; (E) Mellitin. Uniprot entry code: P68727, P56676, P30288, P60514 and P01501, respectively.

Besides peptides, some studies have evaluated the analgesic activity of acylpolyamines that can be used as new alternative drugs for the treatment of chronic pain, as well as tools for the study of the functional role of the AMPA/kainate receptors in the processing of nociceptive pain [77]. In this regard, intrathecal administration of different doses of these toxins blocked thermally induced allodynia [78] and hyperalgesia [79]. The effect of these neurotoxins may suggest a possible involvement of AMPA receptors in the spinal cord during the nociceptive excitatory stimulation [80,81].

4. Antiepileptic and neuroprotective effects

Neurodegenerative disorders comprise a wide range of conditions mostly characterized by a progressive loss of neuronal function and neuronal cell death. The incidence of these diseases in population differs greatly. In conditions such as Parkinson disease and Alzheimer, the number of cases significantly increases in elderly, whereas epileptic patients are mostly children and adolescents. Many processes may trigger neuronal cell death, such as trauma, stroke, tumors, infections, genetic factors and biochemical alterations. Among the latest, the alterations in Ca^{2+}-mediated signaling is thought to play a key role in many neurodegenerative disorders and the increase in intracellular Ca^{2+} concentration might alter neuronal membrane potential [82]. Moreover, the hyperactivation of excitatory transmission mediated mostly by L-glutamate and its ionotropic receptors; kainate, AMPA and NMDA, is responsible for the excessive cationic influx that depolarizes neuronal cells and lead to sustained hyperexcitation observed in brain pathologies such as epilepsy [83]. This increase in glutamatergic activity often referred to as glutamate excitotoxicity [84], might also involve non-receptor neurochemical events such as failure in glutamate uptake system, which ends with an increase in the availability of this neurotransmitter in the synaptic cleft [85,86]. The importance of L-glutamate in neurological disorders relies on the fact that this neurotransmitter is release in the great majority of fast synapses in CNS [84,83]. In this context, many molecules mostly peptides and acylpolyamines, acting on ion channels, receptors and transporters were isolated from arthropod venoms, remarkably spiders, scorpions and wasps [3]. According to [82], polyamines are non-specific antagonist of ligand-gated ion channels, acting at glutamatergic and Ach receptors in an uncompetitive way, that is, the receptor must be activated in order to occur the blockade. This mode of action might diminish the side effects of newly designed medicines, since it blocks only the activated receptors, but does not prevent their opening.

The venom of the orb-web spider *Nephilia clavata* was one of the first venoms studied during the 80s, which resulted in the identification of small compounds named acylpolyamines, among whose we may find jorotoxin (JSTX), one of the first glutamate receptor uncompetitive antagonists [83,84]. Together with JSTX, another polyamines such as argiopin from the venom of the spider *Argiope lobata* [85] and philantotoxin (PhTx) from the venom of the solitary wasp *Philanthus triangulum* [86]. Following the structural characterization and studies in insect or crustaceans, the reports on the action of these polyamines in mammalian CNS started to take place, mostly during the 90s [87]. JSTX-1 and JSTX-3 are synthetic analogues of JSTX. The first inhibits kainate-induced seizures, whereas the latter block glutamate release and hippocampal epileptic discharges [88,89]. Later, JSTX-3 was shown to inhibit the formation of superoxide dismutase-1 (SOD-1) aggregates that lead mutant motor neurons (mSOD-1) to death during the familiar form of the neurodegenerative disease, amyotrophic lateral sclerosis [90]. The authors concluded that increased Ca^{2+} influx mainly through AM-

PA/kainate glutamate receptors make mutant neurons more vulnerable to damage and therefore, JSTX-3 is an interesting neuroprotective agent in this model.

The fraction of the venom of the spider *Agelenopsis aperta* containing argiotoxin, was first demonstrated to have anticonvulsant in NMDA-induced and audiogenic seizures [91]. The synthetic analogue of argiotoxin, Arg-636, is a selective antagonist of NMDA receptors binding to the Mg^{2+} binding site at the receptor with anticonvulsant and neuroprotective actions. In addition, from the venom of *A. aperta*, another NMDA receptor blocker, Agatoxin 489 was reported as anticonvulsant against kainate-induced seizures and its synthetic analogue Agel-505, was able to block cationic currents in oocytes transfected with NMDA receptor cDNA [92].

Aside from antagonizing glutamate receptors, arthropod neurotoxins may exert anticonvulsant and neuroprotective effects targeting other neurotransmitter systems. The venom of the Brazilian spider *Phoneutria nigriventer* has been extensively studied over the past 20 years. Neurotoxins isolated from the venom of *P. nigriventer*, such as PhTx-3 (Tx-3) were reported to inhibit Ca^{2+} dependent-glutamate release [47]. Tx3-3 and Tx3-4 also inhibit voltage-activated Ca^{2+} channels of P/Q type [93] and recently their neuroprotective activity was tested. According to [94], Tx3-3 and Tx3-4 protected hippocampal slices against damage and cell death induced by ischemic insult resulted from low oxygen and low glucose. Moreover, PhTx3, Tx3-3, and Tx3-4, inhibited cell loss in retinal slices submitted to the same ischemic protocol [95]. Another Brazilian species lives in Cerrado, the colonial spider *Parawixia bistriata* and has many neuroactive molecules with different modes of action [96]. Parawixin-1 was the first isolated neurotoxin from *P. bistriata* venom. In experiments using rat retinas, submitted to ischemic insult, the intravitreal injection of Parawixin inhibited cell loss [97], probably through a potent and specific enhancing action on glutamate transporters type EAAT2 [98]. Another neurotoxin isolated from the venom of *P. bistriata*, Parawixin II, formerly, FrPbAII, inhibited GABA and glycine uptakes in synaptosomes from rat cerebral cortices. In addition, the administration of Parawixin II into the vitreous humor of Wistar rats protected retinal neurons against ischemic insult resulted from an increase in the intra ocular pressure [96]. Data also show that Parawixin II blocked seizures induced by the injection of GABAergic antagonists, bicuculline [99], pentylenetetrazole (PTZ) and picrotoxin, as well as pilocarpine and kainic acid [100]. It is worth noting that the acute injection of Parawixin II does not alter rat behavior in the open field and repeated central injection does not impair acquisition and learning in the Morris water maze. Finally, Parawixin II induces ataxia in the rotarod in doses far higher than effective doses, indicating good therapeutic indexes [100].

Also from South America, the Chilean giant pink tarantula *Grammostola spatulata* paralyzes its preys by injecting a mixture of toxins that blocks ion channels [101]. w-Grammotoxin SIA was isolated from the venom of *G. spatulata* and the potent blocking effect over N-, P-, and Q-type but not L-type of voltage gated calcium channels was reported [102]. The antagonistic activity of w-grammotoxin over voltage dependent calcium channels is considered a therapeutic option to be used in neurodegenerative disorders such as ischemia.

The African tarantula *Hysterocrates gigas* known as the giant baboon spider, inhabits the rain forests of West Africa. The isolation of the venom of *H. gigas*, resulted in the identification of the peptide SNX-482 that blocks R-type voltage dependent calcium channels [103].

The arboreal tarantula *Psalmopoeus cambridgei* is an aggressive spider that lives in the tropical forests of Trindad. As mentioned before, PcTx-1 (π-theraphotoxin-Pc1a) present in *P. cambridgei* venom is the only gating modifier of ASICs [50]. In addition to pain inhibitor, it exerts an interesting neuroprotective and a possible antidepressant activity due to the involvement of ASICs in cell excitability. A drop in pH from neutral 7.4 to more acid extracellular environments, lead to opening of ASICs Na^+ or Ca^{2+} permeable pore, membrane depolarization and increase in Ca^{2+} intracellular concentration [104].

In the light of these facts, Yang and coworkers [105] investigated the neuroprotective activity of PcTx in neurons from newborn piglets submitted to a model of asphyxia-induced cardiac arrest. Data show that the administration of PcTx before the hypoxia-ischemia insult partially prevents the death of neurons in putamen, the most vulnerable encephalic area in this model. The addition of MK-801, a NMDA antagonist, in combination with PcTx exerted better results in cell survival, but in low doses of MK-801. In addition to protection of neuronal cells, treatment with PcTx accelerated neurologic recovery. These results point PcTx as a very unique neurotoxin that should be used as tool in the investigation of processes underlying neuroprotection as well as the design of novel neuroprotective agents.

Bees and wasps are part of the group of the insects, whose stings release a cocktail of toxins, including enzymes, peptides and biogenic amines [106]. Toxins in bee venom have received attention for their properties as anti-inflammatory agents, and in many countries, physicians even prescribe bee stings as treatment of rheumatologic diseases. Recently, Doo and colleagues [107] showed that the bee venom when injected in rats with induced Parkinson disease prevent dopamine neurons cell death, possibly by the inhibition of Jun activation.

Regarding solitary wasps, the most studied wasp species is the European beewolf, *Philanthus triangulum*, the natural predators of honeybees. The adult individuals of this species are herbivores, whereas the larvae eat the paralyzed bees brought to the colony by foraging wasps. The isolation of venom contents begun in the early 80s and revealed that among other classes of molecules, *P. triangulum* venom contains potent acylpolyamines [86]. Philanthotoxins, like other acylpolyamines are mostly potent and selective antagonists of vertebrate and invertebrate glutamate receptors, particularly AMPA receptors [108]. The first isolated and most studied philathotoxin is PhTX-433 and its synthetic analogue, PhTx-343, which antagonize Ach and glutamate ionotropic receptors. The neuroprotective activity of PhTx-343 was tested in cerebellar granule cells culture challenged with NMDA and kainate toxicity and compared to that of Arg-636 [109]. Data showed that both polyamines protected cultures against damage, but Arg-636 was found to be less potent than PhTx-343 against kainate-induced damage. The structural change in PhTx-343 increased its potency, but in higher doses, toxic side effects, were observed.

Due to their lack of selectivity, the use of philanthotoxins as pharmaceuticals may have been limited, and so many modified synthetic analogues were designed for medical treatment purposes, so far [82]. However, the use of philanthotoxins and other polyamines as tools in research investigation has aided the understanding of several synaptic mechanisms. As it has been recently shown using Ca^{2+}-permeable AMPA receptors expressed in HEK cells. According to [110] the block of these AMPA receptors by PhTx-74, a synthetic analogue of PhTX-433 will reflect structural and biophysical parameters of the channel, such as its subunit composition and mean conductance, respectively. In addition, the investigation of the antagonistic activity of PhTx-343 over ACh receptors showed that the interaction of the toxin with nicotinic receptors is largely voltage dependent, slow and uncompetitive, a similar mode by which they block glutamate ionotropic channels [111].

Going further on wasp venoms, the anticonvulsant and/or neuroprotective effects of molecules in the venom of two Brazilian social species of the genus *Polybia*, were investigated. According to Cunha and co-workers [112] and Mortari and co-workers [113], the non-enzymatic fraction of the venoms of *Polybia ignobilis* and *Polybia occidentalis* inhibit seizures evoked by the injection of several chemoconvulsants in Wistar rats. The neuroactive molecules present in the venom of *P. ignobilis* and *P. occidentalis* are now in phase of structure-function investigation.

Finally, neurotoxins from scorpion venoms have been subject of a wide range of works, mostly approaching the identification of voltage-dependent ion channel activators/blockers. The neuroprotective and/or anticonvulsant activity of these peptides, in turn have received a few lines of investigation [3] despite the ancient use of these animals whole or parts, in the popular medicine in oriental countries, like China [20]. One of the most studied species is the Asian scorpion *Buthus martensi Karsch* whose venom has several neuroactive peptides, among whose, we may find BmK AEP, which was the first anticonvulsant peptide isolated from scorpion venoms. According to [28], the injection of BmK AEP blocked seizures induced by the injection of coriaria lactone in doses causing no visible side effects [114]. Further isolation of venom of *B. martensi* led to the identification of other peptides, such as BmK AS and BmK Ts and other mostly with analgesic activity. According to Zhao and co-workers [115] BmK AS, a sodium channel modulator at site-4 receptor, inhibited PTZ induced behavioral and electroencephalographic seizures and decreased mean score of pilocarpine-induced seizures. Moreover, these authors showed that BmK AS does not impair locomotion or motor behavior.

The venom of the Mexican scorpion *Centruroides limpidus limpidus*, was fractionated and many activators of voltage-gated ion channel ligands were identified [116]. An exception is Cll9, which stands for *Centruroides limpidus limpidus* toxin nr 9. Cll9 is a 63-residue peptide that has a divergent mode of action; it inhibits sodium channels in superior cervical ganglion neurons and [117]. When injected in Wistar rats via i.c.v., Cll9 inhibited behavioral and electroencephalographic seizures evoked by the microinjection of penicillin into the basolateral amygdala. It is worth noticing that Cll9 has no effect on arthropods such as crickets or crayfish like many sodium channels modulators found in scorpion venoms.

5. Actions on mood disorders

According to the World Health Organization, depression, one of the most important mood disorders, affects up to 5-10% of people worldwide at any time of their lives. Patients with a diagnosed mood disorder are more likely to be women, in productive years, 20 to 40 year-old, and will need in most cases, psychotherapy and/or pharmacological intervention. The costs of these psychiatric and/or psychological disorders are immense, since they affect people regardless of education or socioeconomic status, accounting in the worse cases, for a huge number of suicides. In the United States up to 95% of all suicides, involve mentally ill people, accounting for 1.3% of all deaths [118]. A recent survey shows that generalized anxiety disorder, posttraumatic stress disorder, social anxiety disorder and panic disorder are highly predictive of suicidal idealization [118]. Many aspects of the pathophysiology of mood disorders, as well as the regulation of normal mood states remain unknown. However, with the improvement of techniques for research and diagnosis, such as positron emission tomography, magnetic resonance and multiple channels recording electroencephalogram, soon researchers will be capable to identify structures and neuro-chemical mechanism involved in the regulation of mental states, including mood. So far, we know that limbic structures, such as the amygdala, hippocampus, hypothalamus and pre-frontal cortex control the emotional aspects of brain function. There are plenty of connections among these structures, which might be involved in the onset of mood disorders [119].

Pharmacological treatment of mood disorders consists in daily intake of anti-depressants, anxiolytics or anti-psychotics, most of which cause a wide set of undesired side effects that impose restrictions to patients quality of life. In this regard novel drugs prescribed for mood disorder, such as serotonin uptake inhibitors might be better tolerated and safer than classical drugs, such as monoamine oxidase inhibitors. Among the observed undesired effects we can cite; dizziness, sedation, sexual dysfunction and suicidal though, a paradoxical effect of serotonin uptake inhibitors [119]. Aside from tolerability, medicines used as treatment of depression for example, take too long to produce effect in only a minority of patients; 35-45% of treated patients [120]. Therefore, there is a still great need for novel alternatives to be used both, in basic and clinical science.

The neurochemistry of mood disorders is complex and there is a list of candidates for targets of mood stabilizers, such as adrenaline, GABA, serotonin and glutamate receptors and transporters. There is not many works relating neurotoxins from arthropods and mood disorders, but the few works available showed that in some cases, these molecules might contribute for the development of novel drugs.

The venom of the Brazilian colonial spider *Parawixia bistriata* was fractionated and tested in many animal models of epilepsy, neurodegeneration and anxiety. According to [121], the microinjection of Parawixin 2 (formerly FrPbAII) in the dorsal hippocampus of male Wistar rats increased the time spent in the open arms of the elevated plus maze. Moreover, rats exposed to the light-dark choice apparatus spent more time in the light side of the box, similarly to what observed for diazepam or nipecotic acid, a GABA transporter-1 (GAT-1) inhibitor [121]. In another investigation of *P. bistriata* venom contents, Saidemberg and co-workers

[122] isolated PwTx-I and tested the inhibitory activity of this neurotoxin and its enantiomers on mammalian monoamine oxidases (MAO)-A and -B. According to these authors, PwTx-I, acted as non-competitive inhibitors of MAO-A and MAO-B. MAO metabolizes monoamines dopamine, serotonin and adrenaline, terminating monoaminergic transmission. Inhibitors or MAO (MAOi) have been extensively used as mood stabilizers and currently they have received attention due to their protective activity against age-induced neurodegenerative disorders [123].

Considering alternative targets for mood stabilizers, interesting results were obtained with PcTx, isolated from the venom of the spider *P. cambridgei*, a selective blocker of ASICs. Data showed that both PcTx and amiloride attenuated the stress-induced hyperthermia, whereas only the administration of PcTx increased number of punished crosses measured in the four-plate test. These results indicate that both blockers could attenuate autonomic anxiety parameters, but only PcTx exerted effects on the behavioral anxiety parameters [124].

The aggressive Brazilian social wasp *Agelaia vicina,* builds huge nests where with over a million of individuals. The neurobiological activity of the venom of *A. vicina,* was investigated. Oliveira and colleagues [125] showed that the central injection of the non-enzymatic fraction of the venom induced catalepsy in Wistar compared to the neuroleptic drug haloperidol, a nonselective D2 dopamine antagonist. This effect was reversed by the injection of theophylline or ketamine. The fractionation of the venom led to the identification of two peptides, AvTx-7, mastoparan, and AvTx-8. The investigation of AvTx-8 mode of action in vivo, was performed in a model of panic induction through the activation of GABAergic pathways connecting mesencephalic substantia nigra pars reticulate to superior colliculus [126]. These experiments showed that intranigral microinjection of AvTx-8 inhibited the panic like response induced by the GABAergic blockade of superior colliculus. These effects were similar to those of baclofen, a $GABA_B$ agonist, but differed from the effects of muscimol, a $GABA_A$ agonist. Since post-synaptic $GABA_B$ is a metabotropic receptor complex with a potassium channel, AvTx-8 could act in many different sites that would end in channels opening and hyperpolarization of neuronal membrane.

6. Tools for the study of the functioning of the CNS: learning and memory

Neurotoxins isolated from arthropod are important tools to study of the normal function of the CNS, especially in the structure-function research of the ion channels and the interaction the blockers and modulators in the regulation of the learning and memory (for revision see [127]). In this context, the principal compound used in study of the mechanism of the learning and memory in models of experimental animals is the apamin. Apamin is a short peptide (18 aa) isolated from the venom of honeybee, *Apis mellifera*. It is generally accepted that apamin selectively blocks small conductance calcium-activated potassium channels (SK or K_{ca}), although evidences point to an allosteric modulation of opening rather than the block of the pore [128]. Upon an increase in intracellular calcium, SK channels will open and allow an outward current of potassium ions that is responsible for the hyperpolarization phase of the action potentials. Most studies on structure-function of SK channels were conducted using apamin blockade. The homomeric or heteromeric expression of these channels occurs in higher brain areas such as the neocortex, hippocampus and sub-cortical areas such as thalamus and basal ganglia as well as in cerebellum and brainstem. Substantial data SK channels show the involvement of SK channels in processes of learning and memory, and apamin blockade of SK lead to an increase in cellular excitability, facilitates synaptic plasticity and memory processes run by the hippocampus. In addition, apamin induces alterations in dendritic morphology that might counteract aging and neurodegenerative processes that lead to cognitive and memory impairment [129]. In fact, SK channels co-localize with Ca^{2+}-permeable NMDA receptors in the CA1 region of the hippocampus and the entry of calcium in the cell through these receptors might activate SK that will hyperpolarize membrane. The blockade of SK channels will modulate hippocampal excitability that is essential in memory processes such as long-term potentiation a commonly observed event of synaptic plasticity. Due to its actions, the use of apamin as a tool in research has been consolidated. In addition, the therapeutic use of apamin, in order to maintain hippocampal function and avoid the deleterious effects of aging in memory and cognitive processes have also been proposed [129].

Besides apamin, modulators peptides of the potassium channel isolated from scorpion also have been tested in models of the learning and memory. The good examples are: Charybdotoxin isolated from scorpion *Leiurus quinquestriatus*, Kaliotoxin isolated from *Androctonus mauretanicus* and Iberiotoxin from *Buthus tasmulus*. Charybdotoxin is a potent selective inhibitor of high (large or big) conductance Ca^{2+}-activated potassium channels (KCa1.1, BK, or maxi-K), as well as a Kv1.3 channel [130]. Kaliotoxin is a specific inhibitor of Kv1.1 and Kv1.3 [131] and Iberiotoxin is a selective inhibitor of KCa1.1 channels (formerly BK) [132]. These peptides induced an improvement effect in passive avoidance test and olfactory discrimination task [133,134].

7. Final remarks

The stories of voltage-gated, ligand-gated ion channels and venom toxins are very closely tied. Indeed, the isolation and structural characterization of venom molecules provided a plethora of tools that have been used in the investigation of ion channels structure-function relationships. With the aid of arthropod toxins, remarkably, scorpionic toxins, the characterization of sodium channels was possible. Spider and wasps polyamines, in turn are considered unique ligands of glutamatergic and cholinergic ionotropic receptors. Regarding to peptides and small proteins, arthropod venoms possess an arsenal of these molecules that remain largely unknown and consequently, their pharmacological potential is left unexplored.

Due to the mode of action of neurotoxins, their affinity and selectivity for neuronal structures, many researchers consider them as probes to novel drugs design and development. However, despite of the thousands of patents made with neurotoxins in the past 30 years, very few molecules came to commercialization.

Acknowledgements

The authors thank Msc Juliana Castro e Silva for help in preparing the figure.

Author details

Márcia Renata Mortari[1*] and Alexandra Olimpio Siqueira Cunha[2]

*Address all correspondence to: mmortari@unb.br

1 Department of Physiological Sciences, Institute of Biological Sciences, University of Brasília, Brazil

2 Department of Physiology, FMRP, University of São Paulo, Brazil

References

[1] Zhijian, C., Chao, D., Dahe, J., & Wenxin, L. (2006). The effect of intron location on the splicing of BmKK2 in 293T cells. *J Biochem Mol Toxicol,* 20(3), 127-32.

[2] Estrada, G., Villegas, E., & Corzo, G. (2007). Spider venoms: a rich source of acylpolyamines and peptides as new leads for CNS drugs. *Nat Prod Rep,* 24(1), 145-61.

[3] Mortari, M. R., Cunha, A. O., Ferreira, L. B., & dos, Santos. W. F. (2007). Neurotoxins from invertebrates as anticonvulsants: from basic research to therapeutic application. *Pharmacol Ther.,* 114(2), 171-83.

[4] Rogoza, L. N., Salakhutdinov, N. F., & Tolstikov, G. A. (2005). Polymethyleneamine alkaloids of animal origin. I. Metabolites of marine and microbial organisms]. *Bioorg Khim,* 31(6), 563-77.

[5] WHO. (2007). Neurological Disorders:. *Public Health Challenges.*

[6] Bialer, M., & White, H. S. (2010). Key factors in the discovery and development of new antiepileptic drugs. *Nature reviews Drug discovery,* 9(1), 68-82.

[7] Blier, P. (2010). The well of novel antidepressants: running dry. *J Psychiatry Neurosci,* 35(4), 219-20.

[8] Goudet, C., Chi, C. W., & Tytgat, J. (2002). An overview of toxins and genes from the venom of the Asian scorpion Buthus martensi Karsch. *Toxicon official journal of the International Society on Toxinology,* 40(9), 1239-58.

[9] Cherniack, E. P. (2011). Bugs as drugs, part two: worms, leeches, scorpions, snails, ticks, centipedes, and spiders. *Alternative medicine review : a journal of clinical therapeutic*, 16(1), 50-8.

[10] Ratcliffe, N. A., Mello, C. B., Garcia, E. S., Butt, T. M., & Azambuja, P. (2011). Insect natural products and processes: new treatments for human disease. *Insect Biochem Mol Biol*, 41(10), 747-69.

[11] Monteiro, M. C., Romao, P. R., & Soares, A. M. (2009). Pharmacological perspectives of wasp venom. *Protein and peptide letters*, 16(8), 944-52.

[12] Beleboni, R. D., Pizzo, A. B., Fontana, A. C. K., Carolino, R. D. O. G., Coutinho-Netto, J., & dos, Santos. W. F. (2004). Spider and wasp neurotoxins: pharmacological and biochemical aspects. *Eur J Pharmacol*, 493(1-3), 1-17.

[13] Ashburn, M. A., & Staats, P. S. (1999). Management of chronic pain. *Lancet*, 353(167), 865-869.

[14] Stucky, C. L., Gold, M. S., & Zhang, X. (2001). Mechanisms of pain. *Proc Natl Acad Sci U S A.*, 98(21), 11845-6.

[15] Sindrup, S. H., & Jensen, T. S. (1999). Efficacy of pharmacological treatments of neuropathic pain: an update and effect related to mechanism of drug action. *Pain*, 83(3), 389-400.

[16] Villetti, G., Bergamaschi, M., Bassani, F., Bolzoni, P. T., Maiorino, M., Pietra, C., et al. (2003). Antinociceptive activity of the N-methyl-D-aspartate receptor antagonist N-(2-Indanyl)-glycinamide hydrochloride (CHF3381) in experimental models of inflammatory and neuropathic pain. *J Pharmacol Exp Ther*, 306(2), 804-14.

[17] Ji, Y. H., Mansuelle, P., Terakawa, S., Kopeyan, C., Yanaihara, N., Hsu, K., et al. (1996). Two neurotoxins (BmK I and BmK II) from the venom of the scorpion Buthus martensi Karsch: purification, amino acid sequences and assessment of specific activity. *Toxicon : official journal of the International Society on Toxinology*, 34(9), 987-1001.

[18] Guan, R., Wang, C. G., Wang, M., & Wang, D. C. (2001). A depressant insect toxin with a novel analgesic effect from scorpion Buthus martensii Karsch. *Biochim Biophys Acta*, 1549(1), 9-18.

[19] Rodriguez dela Vega, R. C., & Possani, L. D. (2005). Overview of scorpion toxins specific for channels and related peptides: biodiversity, structure-function relationships and evolution. *Toxicon : official journal of the International Society on Toxinology*, 46(8), 831-44.

[20] Xiong, Y. M., Lan, Z. D., Wang, M., Liu, B., Liu, X. Q., Fei, H., et al. (1999). Molecular characterization of a new excitatory insect neurotoxin with an analgesic effect on mice from the scorpion Buthus martensi Karsch. *Toxicon*, 37(8), 1165-80.

[21] Escoubas, P., Stankiewicz, M., Takaoka, T., Pelhate, M., Romi-Lebrun, R., Wu, F. Q., et al. (2000). Sequence and electrophysiological characterization of two insect-selec-

tive excitatory toxins from the venom of the Chinese scorpion Buthus martensi. *FEBS Lett*, 483(2-3), 175 -80 .

[22] Guan, R. J., Wang, M., Wang, D., & Wang, D. C. (2001). A new insect neurotoxin AngP1 with analgesic effect from the scorpion Buthus martensii Karsch: purification and characterization. *The journal of peptide research official journal of the American Peptide Society*, 58(1), 27-35.

[23] Wang, C. Y., Tan, Z. Y., Chen, B., Zhao, Z. Q., & Ji, Y. H. (2000). Antihyperalgesia effect of BmK IT2, a depressant insect-selective scorpion toxin in rat by peripheral administration. *Brain research bulletin*, 53(3), 335-8.

[24] Tan, Z. Y., Xiao, H., Mao, X., Wang, C. Y., Zhao, Z. Q., & Ji, Y. H. (2001). The inhibitory effects of BmK IT2, a scorpion neurotoxin on rat nociceptive flexion reflex and a possible mechanism for modulating voltage-gated Na(+) channels. *Neuropharmacology*, 40(3), 352-7.

[25] Bai, Z. T., Liu, T., Pang, X. Y., Chai, Z. F., & Ji, Y. H. (2007). Suppression by intrathecal BmK IT2 on rat spontaneous pain behaviors and spinal c-Fos expression induced by formalin. *Brain Res Bull*, 73(4-6), 248 -253 .

[26] Zhang, X. Y., Bai, Z. T., Chai, Z. F., Zhang, J. W., Liu, Y., & Ji, Y. H. (2003). Suppressive effects of BmK IT2 on nociceptive behavior and c-Fos expression in spinal cord induced by formalin. *J Neurosci Res*, 74(1), 167-73.

[27] Li, Y. J., Tan, Z. Y., & Ji, Y. H. (2000). The binding of BmK IT2, a depressant insect-selective scorpion toxin on mammal and insect sodium channels. *Neurosci Res*, 38(3), 257-64.

[28] Wang, Y., Wang, L., Cui, Y., Song, Y. B., Liu, Y. F., Zhang, R., et al. (2011). Purification, characterization and functional expression of a new peptide with an analgesic effect from Chinese scorpion Buthus martensii Karsch (BmK AGP-SYPU1). *Biomed Chromatogr*, 25(7), 801-7.

[29] Ji-H, Y., Li-J, Y., Zhang-W, J., Song-L, B., Yamaki, T., Mochizuki, T., et al. (1999). Covalent structures of BmK AS and BmK AS-1, two novel bioactive polypeptides purified from Chinese scorpion Buthus martensi Karsch. *Toxicon*, 37(3), 519-36.

[30] Lan-D, Z., Dai, L., Zhuo-L, X., Feng-C, J., Xu, K., & Chi-W, C. (1999). Gene cloning and sequencing of BmK AS and BmK AS-1, two novel neurotoxins from the scorpion Buthus martensi Karsch. *Toxicon*, 37(5), 815-23.

[31] Chen, B., & Ji, Y. (2002). Antihyperalgesia effect of BmK AS, a scorpion toxin, in rat by intraplantar injection. *Brain Res*, 952(2), 2322-6.

[32] Chen, J., Feng, X. H., Shi, J., Tan, Z. Y., Bai, Z. T., Liu, T., et al. (2006). The anti-nociceptive effect of BmK AS, a scorpion active polypeptide, and the possible mechanism on specifically modulating voltage-gated Na+ currents in primary afferent neurons. *Peptides*, 27(9), 182-92.

[33] Tan, Z. Y., Mao, X., Xiao, H., Zhao, Z. Q., & Ji, Y. H. (2001). Buthus martensi Karsch agonist of skeletal-muscle RyR-1, a scorpion active polypeptide: antinociceptive effect on rat peripheral nervous system and spinal cord, and inhibition of voltage-gated Na(+) currents in dorsal root ganglion neurons. *Neurosci Lett*, 297(2), 65-8.

[34] Wu, Y., Ji, Y. H., & Shi, Y. L. (2001). Sodium current in NG108-15 cell inhibited by scorpion toxin BmKAS-1 and restored by its specific monoclonal antibodies. *J Nat Toxins*, 10(3), 193-8.

[35] Liu, Y. F., Ma, R. L., Wang, S. L., Duan, Z. Y., Zhang, J. H., Wu, L. J., et al. (2003). Expression of an antitumor-analgesic peptide from the venom of Chinese scorpion Buthus martensii karsch in Escherichia coli. *Protein Expr Purif*, 27(2), 253-8.

[36] Cui, Y., Guo, G. L., Ma, L., Hu, N., Song, Y. B., Liu, Y. F., et al. (2010). Structure and function relationship of toxin from Chinese scorpion Buthus martensii Karsch (BmKAGAP): gaining insight into related sites of analgesic activity. *Peptides*, 31(6), 995-1000.

[37] Cao, Z. Y., Mi, Z. M., Cheng, G. F., Shen, W. Q., Xiao, X., Liu, X. M., et al. (2004). Purification and characterization of a new peptide with analgesic effect from the scorpion Buthus martensi Karch. *J Pept Res*, 64(1), 33-41.

[38] Peng, K., Chen, X. D., & Liang, S. P. (2001). The effect of Huwentoxin-I on Ca(2+) channels in differentiated NG108-15 cells, a patch-clamp study. *Toxicon : official journal of the International Society on Toxinology*, 39(4), 491-8.

[39] Chen, J. Q., Zhang, Y. Q., Dai, J., Luo, Z. M., & Liang, S. P. (2005). Antinociceptive effects of intrathecally administered huwentoxin-I, a selective N-type calcium channel blocker, in the formalin test in conscious rats. *Toxicon : official journal of the International Society on Toxinology*, 45(1), 15-20.

[40] Wen, Tao. Z., Gu, Yang. T., Ying, R., Mao, Cai. W., Lin, L., Chi, Miao. L., et al. (2011). The antinociceptive efficacy of HWTX-I epidurally administered in rheumatoid arthritis rats. *Int J Sports Med*, 32(11), 869-74.

[41] Saez, N. J., Senff, S., Jensen, J. E., Er, S. Y., Herzig, V., Rash, L. D., et al. (2010). Spider-venom peptides as therapeutics. *Toxins*, 2(12), 2851-71.

[42] Murakami, M., Nakagawasai, O., Suzuki, T., Mobarakeh, I. I., Sakurada, Y., Murata, A., et al. (2004). Antinociceptive effect of different types of calcium channel inhibitors and the distribution of various calcium channel alpha 1 subunits in the dorsal horn of spinal cord in mice. *Brain Res*, 1024(1-2), 122 -9.

[43] Rajendra, W., Armugam, A., & Jeyaseelan, K. (2004). Toxins in anti-nociception and anti-inflammation. *Toxinofficial journal of the International Society on Toxinology*, 44(1), 1-17.

[44] Souza, A. H., Ferreira, J., Cordeiro, Mdo. N., Vieira, L. B., De Castro, C. J., Trevisan, G., et al. (2008). Analgesic effect in rodents of native and recombinant Ph alpha 1beta

toxin, a high-voltage-activated calcium channel blocker isolated from armed spider venom. *Pain*, 140(1), 115-26.

[45] Dalmolin, G. D., Silva, C. R., Rigo, F. K., Gomes, G. M., Cordeiro, Mdo. N., Richardson, M., et al. (2011). Antinociceptive effect of Brazilian armed spider venom toxin Tx3-3 in animal models of neuropathic pain. *Pain*, 152(10), 2224-32.

[46] Vieira, L. B., Kushmerick, C., Hildebrand, M. E., Garcia, E., Stea, A., Cordeiro, M. N., et al. (2005). Inhibition of high voltage-activated calcium channels by spider toxin PnTx3-6. *J Pharmacol Exp Ther*, 314(3), 1370-7.

[47] Prado, MA, Guatimosim, C., Gomez, M. V., Diniz, C. R., Cordeiro, M. N., & Romano-Silva, MA. (1996). A novel tool for the investigation of glutamate release from rat cerebrocortical synaptosomes: the toxin Tx from the venom of the spider Phoneutria nigriventer. *Biochem J*, 314(Pt1), 145-50.

[48] De Souza, A. H., Lima, M. C., Drewes, C. C., da, Silva. J. F., Torres, K. C., Pereira, E. M., et al. (2011). Antiallodynic effect and side effects of Phalpha1beta, a neurotoxin from the spider Phoneutria nigriventer: comparison with omega-conotoxin MVIIA and morphine. *Toxicon : official journal of the International Society on Toxinology*, 58(8), 626-33.

[49] Malmberg, A. B., & Yaksh, T. L. (1994). Voltage-sensitive calcium channels in spinal nociceptive processing: blockade of N- and P-type channels inhibits formalin-induced nociception. *J Neurosci*, 14(8), 4882-90.

[50] Saez, N. J., Mobli, M., Bieri, M., Chassagnon, I. R., Malde, A. K., Gamsjaeger, R., et al. (2011). A dynamic pharmacophore drives the interaction between Psalmotoxin-1 and the putative drug target acid-sensing ion channel 1a. Mol Pharmacol Epub 2011/08/10., 80(5), 796-808.

[51] Mc Cleskey, E. W., & Gold, M. S. (1999). Ion channels of nociception. *Annu Rev Physiol*, 61, 835-56.

[52] Waldmann, R., Champigny, G., Lingueglia, E., De Weille, J. R., Heurteaux, C., & Lazdunski, M. (1999). H(+)-gated cation channels. *Ann N Y Acad Sci*, 868, 67-76.

[53] Mazzuca, M., Heurteaux, C., Alloui, A., Diochot, S., Baron, A., Voilley, N., et al. (2007). A tarantula peptide against pain via ASIC1a channels and opioid mechanisms. *Nat Neurosci*, 10(8), 943-5.

[54] Pignataro, G., Simon, R. P., & Xiong, Z. G. (2007). Prolonged activation of ASIC1a and the time window for neuroprotection in cerebral ischaemia. *Braina journal of neurology*, 130(Pt 1), 151 -158 .

[55] Lampe, R. A. (1999). Analgesic peptides from venom of grammostola spatulata and use thereof. http://www.patentgenius.com/patent/5776896.html.

[56]   Grishin, E. V., Savchenko, G. A., Vassilevski, A. A., Korolkova, Y. V., Boychuk, Y. A., Viatchenko-Karpinski, V. Y., et al. (2010). Novel peptide from spider venom inhibits 2X3receptors and inflammatory pain. *Ann Neurol*, 67(5), 680-3.

[57]   Chen, J., & Lariviere, W. R. (2010). The nociceptive and anti-nociceptive effects of bee venom injection and therapy: a double-edged sword. *Prog Neurobiol*, 92(2), 151-83.

[58]   Kwon, Y. B., Kang, M. S., Kim, H. W., Ham, T. W., Yim, Y. K., Jeong, S. H., et al. (2001). Antinociceptive effects of bee venom acupuncture (apipuncture) in rodent animal models: a comparative study of acupoint versus non-acupoint stimulation. *Acupunct Electrother Res*, 26(1-2), 59-68.

[59]   Kwon, Y. B., Kim, J. H., Yoon, J. H., Lee, JD, Han, H. J., Mar, W. C., et al. (2001). The analgesic efficacy of bee venom acupuncture for knee osteoarthritis: a comparative study with needle acupuncture. *Am J Chin Med*, 29(2), 187-99.

[60]   Lee-H, J., Kwon-B, Y., Han-J, H., Mar-C, W., Lee-J, H., Yang-S, I., et al. (2001). Bee Venom Pretreatment Has Both an Antinociceptive and Anti-Inflammatory Effect on Carrageenan-Induced Inflammation. *J Vet Med Sci*, 63(3), 251-9.

[61]   Lee, J. D., Park, H. J., Chae, Y., & Lim, S. (2005). An Overview of Bee Venom Acupuncture in the Treatment of Arthritis. *Evidence-based complementary and alternative medicine : eCAM*, 2(1), 79-84.

[62]   Kwon, Y. B., Kang, M. S., Han, H. J., Beitz, A. J., & Lee, J. H. (2001). Visceral antinociception produced by bee venom stimulation of the Zhongwan acupuncture point in mice: role of alpha(2) adrenoceptors. *Neurosci Lett*, 308(2), 133-7.

[63]   Baek, Y. H., Huh, J. E., Lee, J. D., Choi do, Y., & Park, D. S. (2006). Antinociceptive effect and the mechanism of bee venom acupuncture (Apipuncture) on inflammatory pain in the rat model of collagen-induced arthritis: Mediation by alpha2-Adrenoceptors. *Brain Res*, 073-074, 305-10.

[64]   Son, D. J., Kang, J., Kim, T. J., Song, H. S., Sung, K. J., Yun, Y., et al. (2007). Melittin, a major bioactive component of bee venom toxin, inhibits PDGF receptor beta-tyrosine phosphorylation and downstream intracellular signal transduction in rat aortic vascular smooth muscle cells. *Journal of toxicology and environmental health Part A*, 70(15-16), 1350 -5 .

[65]   Lauterwein, J., Bosch, C., Brown, L. R., & Wuthrich, K. (1979). Physicochemical studies of the protein-lipid interactions in melittin-containing micelles. *Biochim Biophys Acta*, 556(2), 244-64.

[66]   Lavialle, F., Levin, I. W., & Mollay, C. (1980). Interaction of melittin with dimyristoyl phosphatidylcholine liposomes: evidence for boundary lipid by Raman spectroscopy. *Biochim Biophys Acta*, 600(1), 62-71.

[67]   Shier WT. (1979). Activation of high levels of endogenous phospholipase A2 in cultured cells. *Proc Natl Acad Sci U S A*, 76(1), 195-9.

[68] Lad, P. J., & Shier, W. T. (1979). Activation of microsomal guanylate cyclase by a cytotoxic polypeptide: melittin. *Biochem Biophys Res Commun*, 89(1), 315-21.

[69] Park, H. J., Lee, H. J., Choi, M. S., Son, D. J., Song, H. S., Song, M. J., et al. (2008). JNK pathway is involved in the inhibition of inflammatory target gene expression and NF-kappaB activation by melittin. *J Inflamm (Lond).*, 5, 7.

[70] Merlo, L. A., Bastos, L. F., Godin, A. M., Rocha, L. T., Nascimento, E. B. Jr, Paiva, A. L., et al. (2011). Effects induced by Apis mellifera venom and its components in experimental models of nociceptive and inflammatory pain. *Toxicon : official journal of the International Society on Toxinology*, 57(5), 764-71.

[71] Shkenderov, S., & Koburova, K. (1982). Adolapin- A newly isolated analgetic and anti-inflammatory polypeptide from bee venom. *Toxicon*, 20(1), 317-21.

[72] Koburova, K. L. S. G., & Michailova, S. V. (1985). Shkenderov Further investigation on the antiinflammatory properties of adolapin--bee venom polypeptide. *Acta Physiol Pharmacol Bulg*, 11(2), 50-5.

[73] Mortari, M. R., Cunha, A. O., Carolino, R. O., Coutinho-Netto, J., Tomaz, J. C., Lopes, N. P., et al. (2007). Inhibition of acute nociceptive responses in rats after i.c.v. injection of Thr6-bradykinin, isolated from the venom of the social wasp, Polybia occidentalis. *Br J Pharmacol*, 151(6), 860-9.

[74] Pellegrini, M., Mammi, S., Peggion, E., & Mierke, D. F. (1997). Threonine6-bradykinin: structural characterization in the presence of micelles by nuclear magnetic resonance and distance geometry. *J Med Chem*, 40(1), 92-8.

[75] Pellegrini, M., & Mierke, D. F. (1997). Threonine6-bradykinin: molecular dynamics simulations in a biphasic membrane mimetic. *J Med Chem*, 40(1), 99-104.

[76] Nagy, I., Paule, C., White, J., & Urban, L. (2007). Taking the sting out of pain. *Br J Pharmacol*, 151(6), 721-2.

[77] Kawai, N., Shimazaki, K., Sahara, Y., Robinson, H. P., & Nakajima, T. (1991). Spider toxin and the glutamate receptors. *Comp Biochem Physiol C*, 98(1), 87-95.

[78] Sorkin, L. S. Y. T., & Doom, C. M. (1999). Mechanical allodynia in rats is blocked by a Ca2+ permeable AMPA receptor antagonist. *Neuroreport*, 10(17), 3523-6.

[79] Stanfa, L. C., Hampton, D. W., & Dickenson, A. H. (2000). Role of Ca2+-permeable non-NMDA glutamate receptors in spinal nociceptive transmission. Neuroreport Epub 2000/10/24., 11(14), 3199-202.

[80] Jones, M. G., & Lodge, D. (1991). Comparison of some arthropod toxins and toxin fragments as antagonists of excitatory amino acid-induced excitation of rat spinal neurones. *Eur J Pharmacol*, 204(2), 203-9.

[81] Sorkin, L. S., Yaksh, T. L., & Doom, C. M. (2001). Pain models display differential sensitivity to Ca2+-permeable non-NMDA glutamate receptor antagonists. *Anesthesiology*, 95(4), 965-73.

[82] Andersen, T. F., Vogensen, S. B., Jensen, L. S., Knapp, K. M., & Stromgaard, K. (2005). Design and synthesis of labeled analogs of PhTX-56, a potent and selective AMPA receptor antagonist. *Bioorg Med Chem*, 13(17), 5104-12.

[83] Dingledine, R., Borges, K., Bowie, D., & Traynelis, S. F. (1999). The glutamate receptor ion channels. *Pharmacol Rev*, 51(1), 7-61.

[84] Rajendra, W., Armugam, A., & Jeyaseelan, K. (2004). Neuroprotection and peptide toxins. *Brain Res Brain Res Rev*, 45(2), 125-41.

[85] Grishin, E. V., Volkova, T. M., & Arseniev, A. S. (1989). Isolation and structure analysis of components from venom of the spider Argiope lobata. *Toxicon*, 27(5), 541-9.

[86] Clark, R. B., Donaldson, P. L., Gration, K. A., Lambert, J. J., Piek, T., Ramsey, R., et al. (1982). Block of locust muscle glutamate receptors by delta-philanthotoxin occurs after receptor activations. *Brain Res*, 241(1), 105-14.

[87] Usherwood, P. N., & Blagbrough, I. S. (1991). Spider toxins affecting glutamate receptors: polyamines in therapeutic neurochemistry. Pharmacol Ther Epub 1991/11/01., 52(2), 245-68.

[88] Herold, E. E., & Yaksh, T. L. (1992). Anesthesia and muscle relaxation with intrathecal injections of AR636 and AG489, two acylpolyamine spider toxins, in rat. *Anesthesiology*, 77(3), 507-12.

[89] Kanai, H., Ishida, N., Nakajima, T., & Kato, N. (1992). An analogue of Joro spider toxin selectively suppresses hippocampal epileptic discharges induced by quisqualate. *Brain Res*, 581(1), 161-4.

[90] Roy, J., Minotti, S., Dong, L., Figlewicz, D. A., & Durham, H. D. (1998). Glutamate potentiates the toxicity of mutant Cu/Zn-superoxide dismutase in motor neurons by postsynaptic calcium-dependent mechanisms. *J Neurosci*, 18(23), 9673-84.

[91] Seymour, P. A. M. E. (1989). In vivo antagonist activity of the polyamine spider venom component, Argiotoxin-636. *Soc Neurosci Abs*, 15, 463-24.

[92] K, W. (1993). Effects of Agelenopsis aperta toxins on the N-methyl-D-aspartate receptor: polyamine-like and high-affinity antagonist actions. *J Pharmacol Exp Ther*, 266(1), 231-6.

[93] Miranda, D. M., Romano-Silva, M. A., Kalapothakis, E., Diniz, C. R., Cordeiro, M. N., Moraes-Santos, T., et al. (2001). Spider neurotoxins block the beta scorpion toxin-induced calcium uptake in rat brain cortical synaptosomes. *Brain Res Bull*, 54(5), 533-6.

[94] Pinheiro, A. C., da Silva, A. J., Prado Cordeiro, MA, Mdo, N., Richardson, M., Batista, M. C., et al. (2009). Phoneutria spider toxins block ischemia-induced glutamate release, neuronal death, and loss of neurotransmission in hippocampus. *Hippocampus*, 19(11), 1123-9.

[95] Agostini, R. M., Nascimento do Pinheiro, A. C., Binda, N. S., Romano, Silva., Nascimento do, Cordeiro. M., Richardson, M., et al. (2011). Phoneutria spider toxins block

ischemia-induced glutamate release and neuronal death of cell layers of the retina. *Retina*, 31(7), 1392-9.

[96] Beleboni, R. O., Guizzo, R., Fontana, A. C., Pizzo, A. B., Carolino, R. O., Gobbo-Neto, L., et al. (2006). Neurochemical characterization of a neuroprotective compound from Parawixia bistriata spider venom that inhibits synaptosomal uptake of GABA and glycine. *Mol Pharmacol*, 69(6), 1998-2006.

[97] Fontana, A. C., Guizzo, R., de Oliveira, Beleboni. R., Meirelles, E. S. A. R., Coimbra, N. C., Amara, S. G., et al. (2003). Purification of a neuroprotective component of Parawixia bistriata spider venom that enhances glutamate uptake. *Br J Pharmacol*, 139(7), 1297-309.

[98] Fontana, A. C., de Oliveira, Beleboni. R., Wojewodzic, M. W., Ferreira, Dos., Santos, W., Coutinho-Netto, J., Grutle, N. J., et al. (2007). Enhancing glutamate transport: mechanism of action of Parawixin1, a neuroprotective compound from Parawixia bistriata spider venom. *Mol Pharmacol*, 72(5), 1228-37.

[99] Cairrão, M. A. R. R. A., Pizzo, A. B., Fontana, A. C. K., Beleboni, R. O., Coutinho-Neto, J., Miranda, A., & Santos, W. F. (2002). Anticonvulsant and GABA uptake inhibition properties of P. bistriata and S. raptoria spider venom fractions. *Pharm Biol*, 40(6), 472-7.

[100] Gelfuso, E. A., Cunha, A. O., Mortari, M. R., Liberato, J. L., Paraventi, K. H., Beleboni, R. O., et al. (2007). Neuropharmacological profile of FrPbAII, purified from the venom of the social spider Parawixia bistriata (Araneae, Araneidae), in Wistar rats. *Life Sci*, 80(6), 566-72.

[101] Lampe, R. A., Defeo, P. A., Davison, Young. J., Herman, J. L., Spreen, R. C., et al. (1993). Isolation and pharmacological characterization of omega-grammotoxin SIA, a novel peptide inhibitor of neuronal voltage-sensitive calcium channel responses. *Mol Pharmacol*, 44(2), 451-60.

[102] Piser, T. M., Lampe, R. A., Keith, R. A., & Thayer, S. A. (1995). Complete and reversible block by ω-grammotoxin SIA of glutamatergic synaptic transmission between cultured rat hippocampal neurons. *Neurosci Lett*, 201(2), 135-8.

[103] Newcomb, R. P. A., Szoke, B. G., Tarczy-Hornoch, K., Hopkins, W. F., Cong, R. L., Miljanich, G. P., Dean, R., Nadasdi, L., Urge, L., & Bowersox, S. S. (2005). inventor; Elan Pharmaceuticals Inc US, assignee. Class e voltage-gated calcium channel antagonist and methods patent 0920504

[104] Chu, X. P., Papasian, C. J., Wang, . J. Q., & Xiong, Z. G. (2011). Modulation of acid-sensing ion channels: molecular mechanisms and therapeutic potential. *International journal of physiology, pathophysiology and pharmacology*, 3(4), 288-309.

[105] Yang, Z. J., Ni, X., Carter, E. L., Kibler, K., Martin, L. J., & Koehler, R. C. (2011). Neuroprotective effect of acid-sensing ion channel inhibitor psalmotoxin-1 after hypoxia-ischemia in newborn piglet striatum. *Neurobiol Dis*, 43(2), 446-54.

[106] Libersat, F. (2003). Wasp uses venom cocktail to manipulate the behavior of its cockroach prey. *Journal of comparative physiology A, Neuroethology, sensory, neural, and behavioral physiology*, 189(7), 497-508.

[107] Doo, A. R., Kim, S. T., Kim, S. N., Moon, W., Yin, C. S., Chae, Y., et al. (2010). Neuroprotective effects of bee venom pharmaceutical acupuncture in acute 1-methyl-4-phenyl-1,2,3,6-tetrahydropyridine-induced mouse model of Parkinson's disease. *Neurol Res*, 32(1), 88-91.

[108] Stromgaard, K. (2005). Natural products as tools for studies of ligand-gated ion channels. *Chem Rec*, 5(4), 229-39.

[109] Green, A. C., Nakanishi, K., & Usherwood, P. N. (1996). Polyamine amides are neuroprotective in cerebellar granule cell cultures challenged with excitatory amino acids. *Brain Res*, 717(1-2), 135 -46 .

[110] Jackson, A. C., Milstein, A. D., Soto, D., Farrant, M., Cull-Candy, S. G., & Nicoll, R. A. (2011). Probing TARP modulation of AMPA receptor conductance with polyamine toxins. *J Neurosci*, 31(20), 7511-20.

[111] Brier, T. J., Mellor, I. R., Tikhonov, D. B., Neagoe, I., Shao, Z., Brierley, MJ, et al. (2003). Contrasting actions of philanthotoxin-343 and philanthotoxin-(12) on human muscle nicotinic acetylcholine receptors. *Mol Pharmacol*, 64(4), 954-64.

[112] Cunha, A. O., Mortari, M. R., Oliveira, L., Carolino, R. O., Coutinho-Netto, J., & dos Santos, W. F. (2005). Anticonvulsant effects of the wasp Polybia ignobilis venom on chemically induced seizures and action on GABA and glutamate receptors. *Comparative biochemistry and physiology Toxicology & pharmacology : CBP*, 141(1), 50-7.

[113] Mortari, M. R., Cunha, A. O., de Oliveira, L., Vieira, E. B., Gelfuso, E. A., Coutinho-Netto, J., et al. (2005). Anticonvulsant and behavioural effects of the denatured venom of the social wasp Polybia occidentalis (Polistinae, Vespidae). *Basic & clinical pharmacology & toxicology*, 97(5), 289-95.

[114] Wang, C. G., He, X. L., Shao, F., Liu, W., Ling, M. H., Wang, D. C., et al. (2001). Molecular characterization of an anti-epilepsy peptide from the scorpion Buthus martensi Karsch. *Eur J Biochem*, 2268(8), 480-5.

[115] Zhao, R., Weng-C, C., Feng, Q., Chen, L., Zhang-Y, X., Zhu-Y, H., et al. (2011). Anticonvulsant activity of BmK AS, a sodium channel site 4-specific modulator. *Epilepsy & Behavior*, 20(2), 267-76.

[116] Martin, BM R. A., Gurrola, G. B., Nobile, M., Prestipino, G., Possani, L. D., & Novel, K. (1994). channel-blocking toxins from the venom of the scorpion Centruroides limpidus limpidusKarsch. *Biochem J*, 15(304), 51-6.

[117] Corona, M., Coronas, F. V., Merino, E., Becerril, B., Gutiérrez, R., Rebolledo-Antunez, S., et al. (2003). A novel class of peptide found in scorpion venom with neurodepressant effects in peripheral and central nervous system of the rat. *Biochimica et Biophysica Acta BBA)- Proteins & Proteomics*, 1649(1), 58-67.

[118] Cougle, J. R., Keough, Riccardi. C. J., & Sachs-Ericsson, N. (2009). Anxiety disorders and suicidality in the National Comorbidity Survey-Replication. *J Psychiatr Res*, 43(9), 825-9.

[119] Hatcher, S., & Arroll, B. (2012). Newer antidepressants for the treatment of depression in adults. *BMJ*, 344, d8300.

[120] Nemeroff, C. B., & Owens, M. J. (2002). Treatment of mood disorders. *Nat Neurosci. Suppl*, 5, 1068-70.

[121] Liberato, J. L., Cunha, A. O., Mortari, M. R., Gelfuso, E. A., Beleboni, Rde. O., Coutinho-Netto, J., et al. (2006). Anticonvulsant and anxiolytic activity of FrPbAII, a novel GABA uptake inhibitor isolated from the venom of the social spider Parawixia bistriata (Araneidae: Araneae). *Brain Res*, 1124(1), 19-27.

[122] Saidemberg, D., Ferreira, M., Takahashi, M. A., Gomes, T. N., Cesar-Tognoli, P. C., da, L. M., Silva-Filho, L. C., et al. (2009). Monoamine oxidase inhibitory activities of indolylalkaloid toxins from the venom of the colonial spider Parawixia bistriata: functional characterization of PwTX-I. *Toxicon*, 54(6), 717-24.

[123] Maruyama, M., & Na, W. Monoamine Oxidase Inhibitors as Neuroprotective Agents in Age-Dependent Neurodegenerative Disorders. *Curr Pharm Des*, 19, 799-817.

[124] Dwyer, J. M., Rizzo, S. J., Neal, S. J., Lin, Q., Jow, F., Arias, R. L., et al. (2009). Acid sensing ion channel (ASIC) inhibitors exhibit anxiolytic-like activity in preclinical pharmacological models. *Psychopharmacology (Berl)*, 203(1), 41-52.

[125] de Oliveira, L., Cunha, A. O., Mortari, M. R., Coimbra, N. C., & Dos, Santos. W. F. (2006). Cataleptic activity of the denatured venom of the social wasp Agelaia vicina (Hymenoptera, Vespidae) in Rattus norvegicus (Rodentia, Muridae). *Prog Neuropsychopharmacol B ol Psychiatry*, 30(2), 198-203.

[126] de Oliveira, L., Cunha, A. O., Mortari, M. R., Pizzo, A. B., Miranda, A., Coimbra, N. C., et al. (2005). Effects of microinjections of neurotoxin AvTx8, isolated from the social wasp Agelaia vicina (Hymenoptera, Vespidae) venom, on GABAergic nigrotectal pathways. *Brain Res*, 1031(1), 74-81.

[127] Gati, C. D., Mortari, M. R., & Schwartz, E. F. (2012). Towards therapeutic applications of arthropod venom k(+)-channel blockers in CNS neurologic diseases involving memory acquisition and storage. *Journal of toxicology*, 756358.

[128] Dilly, S., Lamy, C., Marrion, N. V., Liegeois, . J. F., & Seutin, . V. (2011). Ion-channel modulators: more diversity than previously thought. *Chembiochem : a European journal of chemical biology*, 12(12), 1808-12.

[129] Romero-Curiel, A., Lopez-Carpinteyro, D., Gamboa, C., De la Cruz, F., Zamudio, S., & Flores, G. (2011). Apamin induces plastic changes in hippocampal neurons in senile Sprague-Dawley rats. *Synapse*, 65(10), 1062-72.

[130] Gimenez-Gallego, G., Navia, MA, Reuben, J. P., Katz, G. M., Kaczorowski, G. J., & Garcia, M. L. (1988). Purification, sequence, and model structure of charybdotoxin, a

potent selective inhibitor of calcium-activated potassium channels. *Proc Natl Acad Sci U S A*, 85(10), 3329-33.

[131] Crest, M., Jacquet, G., Gola, M., Zerrouk, H., Benslimane, A., Rochat, H., et al. (1992). Kaliotoxin, a novel peptidyl inhibitor of neuronal BK-type Ca(2+)-activated K+ channels characterized from Androctonus mauretanicus mauretanicus venom. *J Biol Chem*, 267(3), 1640-7.

[132] Galvez, A., Gimenez-Gallego, G., Reuben, J. P., Roy-Contancin, L., Feigenbaum, P., Kaczorowski, G. J., et al. (1990). Purification and characterization of a unique, potent, peptidyl probe for the high conductance calcium-activated potassium channel from venom of the scorpion Buthus tamulus. *J Biol Chem*, 265(19), 11083-90.

[133] Kourrich, S., Mourre, C., & Soumireu-Mourat, B. (2001). Kaliotoxin, a Kv1.1 and Kv1.3 channel blocker, improves associative learning in rats. *Behav Brain Res*, 120(1), 35-46.

[134] Edwards, T. M., & Rickard, N. S. (2006). Pharmaco-behavioural evidence indicating a complex role for ryanodine receptor calcium release channels in memory processing for a passive avoidance task. *Neurobiol Learn Mem*, 86(1), 1-8.

Chapter 5

Toxins from *Lonomia obliqua* — Recombinant Production and Molecular Approach

Ana Marisa Chudzinski-Tavassi,
Miryam Paola Alvarez-Flores,
Linda Christian Carrijo-Carvalho and
Maria Esther Ricci-Silva

Additional information is available at the end of the chapter

1. Introduction

Few species of butterflies and moths (order Lepidoptera) are involved in human envenoming [1]. Caterpillars are the larval forms of moths and butterflies. Toxins are usually found in the caterpillar's hairs and spines with defense purposes. The majority of medically important encounters with lepidopterans occur with exposure to the caterpillar's urticating hairs or spines, but hemolymph can also have toxic properties [1, 2]. A variety of clinical effects have been described, which depend on the family and species involved, ranging from local to systemic reactions [3, 4].

In most occasions, the adverse effects caused by caterpillars are self-limited and can be treated with topical antipruritics [4]. However, for the envenoming by the South American *Lonomia obliqua* caterpillars (Figure 1), named lonomism, the antilonomic serum produced at the Butantan Institute in Brazil is the only effective treatment to reestablish the coagulation parameters in poisoned patients and to avoid the complications seen in severe cases such as intracerebral hemorrhage and acute renal failure [5-10].

In 1989, an outbreak of accidents with this species became a serious public health threat in southern Brazil, with high fatality rates [5, 11-15]. Since then, many studies have been carried out to understand the pathophysiological mechanisms of envenoming [14] and to identify the toxins responsible for adverse reactions.

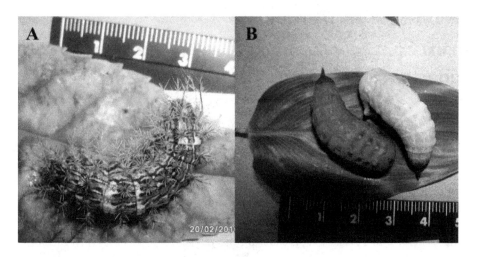

Figure 1. A) *Lonomia obliqua* caterpillar (5th instar) and B) pupa.

L. obliqua is the caterpillar that has the most studied venom, which main components have been isolated and characterized [14, 16, 17]. Table 1 lists the biological activities and toxins isolated and characterized from the bristle extract or hemolymph of *L. obliqua*. *In vivo* studies reported an antithrombotic effect caused by the bristle extract, while most *in vitro* studies reported procoagulant activities [14, 16-23]. It is well known, for a wide range of animal venoms, that procoagulant toxins can cause *in vivo* activation of the coagulation system. The hemostatic disturbances observed in the envenoming by *L. obliqua* caterpillars, result in a consumption coagulopathy (resembling a disseminated intravascular coagulation) and secondary fibrinolysis, which can lead to the hemorrhagic syndrome [6].

The principal components in the caterpillar's venom have been initially identified by isolating toxins through classical purification methods and following the main activities observed in the whole bristle extract (Figure 2). However, this approach provides knowledge restricted only to the most abundant toxins, and usually reveals that activities which are directly associated to the main symptoms and effects observed in the envenoming outcomes. Experimental assays were specifically developed to test the hemostatic and enzymatic activities of *Lonomia* toxins and their actions on the coagulation cascade. This knowledge has been valuable for description and management of the envenoming syndrome, but with the classical approach, low abundant components and unexpected effects are usually overlooked. Possible interaction of venom components, cross-reactions and secondary effects, useful to provide a systemic view of the pathways involved in the toxin's effects are often unnoticed.

In the last years, methods applied in genomic, transcriptomic and proteomic analyses have been applied with the aims of cataloging and classifying the toxins based on their structure and activity (Figure 3). Thus, it was possible to analyze the envenoming processes at the molecular level. For example, significant advance was achieved through two independent transcriptome studies, which generated a list of putative toxic proteins from *L. obliqua* bristles

Activity (toxin)	Source	MW (kDa)	Characteristics and observed effects	Reference
Prothrombin activation (Lopap)	Bristle extract	21	Serine protease, activity increased by Ca^{2+}; consumption coagulopathy *in vivo;* cell survival in endothelial cell culture. Recombinant form produced in bacteria and yeast.	[18, 19, 21, 24-26]
FXa-like	Bristle extract	21	Hydrolytic activity on S-2222 chromogenic substrate, Ca^{2+}-independent; N-terminal sequence similar to Lopap.	[27]
Factor X activation (Losac)	Bristle extract	45	Serine protease, Ca^{2+}-independent; Cell survival in HUVEC. Recombinant form produced in bacteria.	[20-22, 28]
Phospholipase A_2-like	Bristle extract	15	Indirect hemolytic activity in human and rat red blood cells *in vitro*, Ca^{2+}-independent; intravascular hemolysis *in vivo*.	[29-31]
Fibrinogenolytic (Lonofibrase)	Hemolymph	35	αβ fibrinogenase activity; enable to affect fibrin cross-linked.	[32-34]
Hyaluronidase (Lonoglyases)	Bristle extract	49 53	β-endohexosaminidase activity; degradation of extracellular matrix.	[35]
Antiapoptotic	Hemolymph	51	Activity on *Spodoptera frugiperda* (Sf-9) cell culture.	[36]
Antiviral	Hemolymph	20	Antiviral activity against measles, influenza and polio viruses. Recombinant form produced in baculovirus/insect.	[37, 38]
Nociceptive and Edematogenic	Bristle extract	NI	Nociception facilitated by prostaglandin production; edematogenic response facilitated by prostanoids and histamine.	[39]
Kallikrein-kinin system activation	Bristle extract	NI	Kinin release from low molecular weight kininogen; edema formation and fall in arterial pressure.	[40]
Platelet adhesion and aggregation	Bristle extract	NI	Direct platelet aggregation and ATP secretion; activity inhibited by *p*-bromophenacyl bromide, a specific PLA_2 inhibitor.	[41, 42]

NI: None isolated, studies carried out using the whole venom

Table 1. Toxins and activities described in *L. obliqua* venom.

and hemolymph [20, 43]. In addition, significant advance was achieved as a result of micro-array study [44]. Moreover, by coupling proteomics and immunochemical approaches, some immunogenic components were identifying in the bristle extract, especially those related to hemostasis [9]. These components were detected by the antilonomic hyperimmune serum produced at the Butantan Institute, and abundant proteins were identified.

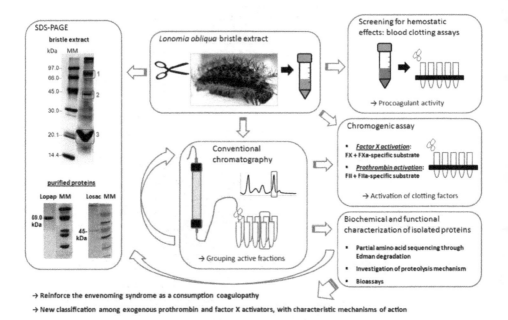

Figure 2. Schematic representation of the classical strategy employed in initial studies of the *Lonomia obliqua* venom. The bristle extract was analyzed through denaturing electrophoresis (SDS-PAGE) which showed the venom is a complex mixture of proteins. Screening assays were carried out to investigate possible effects on blood coagulation and fibrinolysis. The venom showed procoagulant activity by decreasing blood clotting time. Two procoagulant components (Lopap, a prothrombin activator and Losac, a factor X activator) were identified and isolated from the bristle extract for further characterization. The specific activity of Lopap and Losac were observed on purified coagulation factor zymogens (FII or FX) using chromogenic substrates to detect generation of active forms of clotting factors (FIIa and FXa) by these toxins. This assay was used in the purification process to identify the active fractions containing each toxin. SDS-PAGE profile shows Lopap (1- multimer, 3- monomer) and Losac (2) are abundant components in the venom. MM: molecular markers.

Production of recombinant forms of *Lonomia* toxins and discovery of new molecules are opening perspectives in the scientific area for basic and applied researches. These molecules can point out novel mechanisms of action, undiscovered molecular interactions and new classes of enzymes and inhibitors. Interesting, some venom toxins have shown multifunctional properties [19, 22, 28]. The best examples are Lopap (a prothrombin activator with high similarity with lipocalins) and Losac (a factor X activator highly similar to hemolins). Besides activation of blood coagulation, Lopap and Losac can modulate cellular functions and promote cell survival [22, 45]. Both molecules were cloned and produced in its recombinant form in yeast and/or bacteria [19, 25, 28].

Additional studies will be conducted to determine the involvement of the venom components in the envenoming syndrome and their biological significance for physiological processes of the animal, such as insect metamorphosis, which is a combination of growth/activation/ differentiation/programmed cell death signals. Thus, this chapter reviews the currently available information about *L. obliqua* venom, and focus on strategies to unveil molecular aspects of toxins and the perspectives for therapeutic and biotechnological applications.

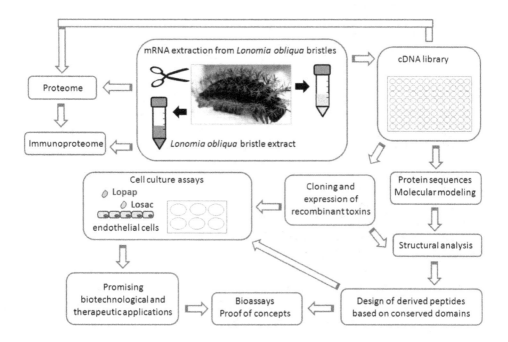

Figure 3. Schematic representation of the strategies to explore the Lonomia obliqua venom and toxins based on cellular and molecular approaches. Results obtained indicate promising applications for these proteins and derived peptides.

2. Molecular approach

For many years, direct purification of toxins from venoms was the best procedure to characterize them with regard to their primary structure. Then, the development of molecular approaches to characterize toxin genes represented an expansion in the understanding of the structure and function of toxin, critical for the development of new treatments directed against the venom toxins (antivenoms). Cloning of cDNAs coding for biochemically isolated toxins has improved their characterization. *Transcriptomic* allows the identification of cellular transcripts in a given cell population, while proteomic studies protein's properties and functions (expression level, structure, post-translational modification, etc.) of proteins expressed by the genome of an organism at a certain point of time. The availability of technologies for high throughput analysis has led to integrate toxin expression at mRNA and protein level. This flow of genetic information from DNA to proteins is the base of the central dogma of molecular biology [46].

2.1. Transcriptomics of *Lonomia obliqua* bristle extract

Expressed Sequence Tags strategy is an approach to characterize the transcriptome of a cell, gland or organism and is based in all the transcript (the most abundant are the mRNAs) produced at a specific time and fully sequenced to create a representative catalogue of expressed genes [47]. Hundreds to a thousand of sequences are grouped into *contigs* or clusters based on DNA sequence information and bioinformatics analysis (Figure 4). Nowadays, the EST-based strategy is commonly employed for identifying expressed genes in species of interest [48, 49]. This approach has been successfully used to compile a lists of genes expressed in venom's glands of a wide range of animals [50-53].

EST-strategy was used to identify the major transcripts present in *L. obliqua* bristle extract [19, 20, 43]. About 702 clusters (representing 1,278 independent clones) were assembled and characterized as lipocalins, hemolins, serine proteases, serine protease inhibitors, serpins, tumor suppressors, ribosomal, structural and cell cycle proteins as shown in Table 2 [20]. Most of the transcripts represent proteins involved in the animal physiology. Those sequences were deposited in data bank (NCBI GenBank accession numbers: CX815710–CX817210, CX820335–CX820336, AY908986) [20]. A pool of DNA sequences showed no similarities with well-known sequences in data bank. The most abundant toxin was a lipocalin of 21 kDa, and analysis of its N-terminal sequence shows 100% homology with Lopap (GenPept accession number: AAW88441). The Lopap whole sequence (accounting for 1.6% of the total clones) was identified in this cDNA library (accession number: AY908986).

Functional categories	No. of clusters	No. of clones	Clones/ clusters	% of Total	% of Hits
General Metabolism	72	94	1.30	6.1	7.0
Transcriptional and translational	165	462	2.80	30.7	37.0
Processing and sorting	10	13	1.30	0.8	1.0
Degradation	9	20	2.22	1.3	2.0
Structural functions	47	243	5.17	16.2	19.0
Cell regulation	26	82	3.15	5.4	6.0
Other functions	138	244	1.77	16.2	19.0
Conserved unknown proteins	42	120	2.86	8.0	9.0
TOTAL	509	1278	2.51	85.0	100.0

Table 2. Major transcripts present in the *Lonomia obliqua* bristle extract identified by EST-based strategy. Adapted from EST data-bank of NCBI deposited by Reis and colleagues in 2004 [19, 20].

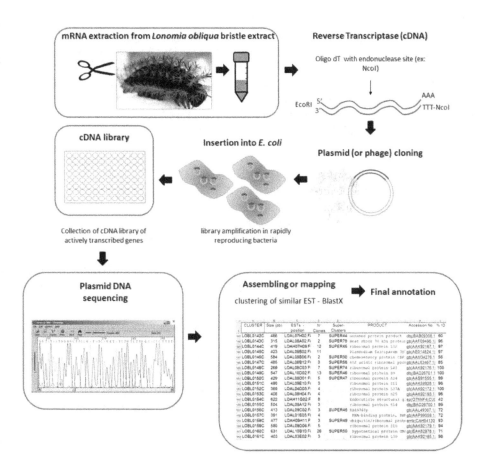

Figure 4. Schematic view of a transcriptomic approach based on EST-strategy. The strategy of construction of the cDNA library starts by the extraction of total RNA from a sample (ex. *L. obliqua* bristle extract). After purification of the mRNAs with an oligo (dT)-cellulose column, the cDNAs are synthesized (reverse transcription) by using synthetic oligo-nucleotides containing a restriction sites (in figure: NcoI and EcoRI for sense and antisense primers, respectively). The cDNA obtained can be inserted into a vector (plasmids or phages), generating a cDNA library. The library can be per-petuated by transforming the clones (plasmids) or infecting themselves (in the cases of phages) in *E. coli*. Based on DNA sequence information, bioinformatics tools predict the amino acid sequence of the corresponding gene products and their similarity to known genes. The redundant EST data sets are organized and integrated into cluster [54, 55].

Other cDNA libraries were constructed from bristles and integument [43]. The transcripts of those libraries revealed the presence of sequences related to trypsin-like enzymes, blood coagulation factors, prophenoloxidase cascade activators, cysteine proteases, phospholipase A2, serpins, cystatins, antibacterial proteins, lipocalins, and others (GenBank accession number: AY829732–AY829859) [43]. Sequences deposited independently in gene banks from both cDNA libraries are complementary. Apart from new venom component precursors, both libraries describe gene products related to cellular processes important for venom production, including high protein synthesis, tuned post-translational processing and trafficking. Those important projects contributed significantly to the characterization of this venom, which showed to be a rich source of proteins and active principles. Further studies about the biological and pharmacological properties of these molecules are necessary to understand its involvement in the envenoming process. Recently, the next-generation of sequencing methods - for example, pyrosequencing - have improved and increased the sequencing reducing time and cost compared to the traditional Sanger method [47, 56].

2.2. Microarray analysis

The identification of genes expressed in cells of a tissue is a basic step to provide essential information about gene function and tissue physiology. The gene expression analysis through the microarray technology (cDNA arrays) has become a powerful tool for rapid analysis of the functional effects of toxins on cells and tissues [57]. The main application of cDNA arrays is to compare the expression of known genes in different physiological situations, for example, tissues in normal and pathological conditions [58]. Thus, analyses of array data contribute to a better understanding of complex gene expression patterns related to physiology and metabolism, unveiling networks or pathways previously unknown.

A study of the effects of L. obliqua bristle extract on the gene expression profile of cultured human fibroblasts showed that many genes are up- and down-regulated, especially those related to the inflammatory processes such as IL-8, IL-6, CXCL1 and CCL2 [44]. Other changes in the expression pattern of some genes, such as prostaglandin-endoperoxide synthase 2, urokinase-type plasminogen activator receptor and tissue factor, were also observed, which could contribute to the pathological effects of lonomism. The authors suggest that the clinical manifestations may be a result of the direct action of L. obliqua venom on the host cells allied to an indirect effect caused by alteration in the gene expression pattern in host tissues.

2.3. Immunoproteome of *Lonomia obliqua* bristle extract

The identification of antigens eliciting an immune response by applying proteomic technologies can be defined as *Immunoproteomics*. Some usual immunoproteomics approaches are shown in Figure 5. Here, the perspective for its application regards the improvement of serum therapy by the selection of antigens for toxin-specific immunization of horses. Furthermore, some applications correlate the identification of antigens with certain diseases, such as infectious, autoimmune or cancer, providing diagnostic and monitoring informations. In this way, these methodologies are good choices in developing clinical applications and also to discover biomarkers. [59].

In classical gel-based strategy, the isolation and identification of proteins/antigens comprises a combination of bidimensional electrophoresis, immunoblotting and mass spectrometry. The aim of bidimensional electrophoresis is to isolate proteins based on their charge and mass [60, 61]. The first step is isoelectric focusing (IEF), where proteins migrate to reach their isoelectric point in an immobilized pH gradient gel under high voltage. All proteins are given negative charge by addition of SDS detergent. This step also includes denaturation of proteins by reduction and alkylation. The second step is SDS-polyacrylamide gel electrophoresis (SDS-PAGE), where smaller proteins migrate faster through the gel to the anode than larger ones. Detection of proteins can be performed by gel staining or immunoblotting.

Immunoblotting involves the transfer of proteins from gel to a nitrocellulose or PVDF membrane in an electric field [62]. The immobilized proteins in the membrane are subsequently incubated with antibodies that have affinity for the proteins of interest. Detection is carried out by enzyme-labelled secondary antibodies against the constant region of the primary IgG antibody, followed by the addition of a chemilluminescent substrate. The substrate reaction can be visualized by fluorescence.

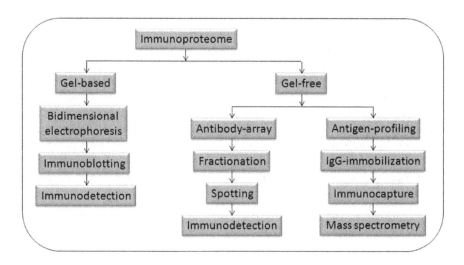

Figure 5. General approaches in Immunoproteomics. **Gel-based approach**: Bidimensional electrophoresis is based on protein separation on their pI and molecular mass. Then, the proteins are transferred from gel and immobilized on membrane (Western blotting). Antigens will be detected after serum incubation, followed by addition of secondary labelled antibodies and their substrates. **Gel-free approaches**: Antigen array: Proteins are fractionated (pI, hydrophobicity,etc) and spotted in a solid support. After that, antigenic fractions can be detected using patient serum and secondary labelled antibodies. Antigen profiling: Immunocapture is based on immobilization of patient immunoglobulins G, which are directly used to capture and isolate antigenic proteins from a complex mixture of proteins. Captured antigens are profiled by mass spectrometry (modified from 66).

Following this, the immunogenic proteins are removed from the gel and enzymatically digested for further mass spectrometry analyses [63]. Trypsin is generally used, cleaving an amide bond on the C-terminal side of lysine and arginine residues, which will be protonated and analyzed in positive-ion mode. Addition of diluted acid (0.1% formic acid or 1% trifluoracetic acid) to the sample contributes to the ionization process.

The ionization methods that are most often used for peptides and proteins are Matrix Assisted Laser Desorption Ionization (MALDI) and Electrospray Ionization (ESI). Peptides and proteins can be identified by Peptide mass fingerprinting (PMF) or *de novo* sequencing [63]. PMF is based on their fragmentation pattern, considering that identical peptide maps have identical amino acid sequences. For *de novo* sequencing analyses, the precursor ion is selected for fragmentation and the product ions are evaluated by mass differences between successive peaks in the spectrum, which are related to the individual mass of their residues.

The identification of immunogenic compounds from *L. obliqua*'s bristle extract was performed on gel-based approach using the polyclonal horse anti-Lonomic hyperimmune serum and anti-Lopap specific rabbit serum produced by the Butantan Institute [9]. Bidimensional electrophoresis of bristle extract revealed 157 silver stained spots, under non-reducing conditions (without DTT and iodoacetamide addition), providing an overview protein mapping (Figure 6A). However, 153 spots were immunodetected using anti-Lonomic serum (Figure 6B) and 30 spots detected using anti-Lopap serum (Figure 6C). Abundant proteins from 24 selected colloidal Coomassie Blue gel spots, corresponding to immunogenic proteins, were digested with trypsin and analysed by tandem mass spectrometry. The identification searches were carried out using the *L. obliqua* bristle EST databank. Lipocalins (spots 05, 09, 10, 14, 15, 16, 18, 24), cuticle protein (spots 05, 06, 07, 08, 11, 12, 13) and serpins (spot 21) were amongst the proteins identified (Figure 6A) [9]. Lipocalins can play a role in homeostasis and inflammation, as a defense mechanism in haematophagous arthropods. Lopap, characterized as a lipocalin protein member, and its all isoforms were highly represented as immunogenic proteins, revealed by the specific anti-Lopap serum (Figure 6C). The bristle' cDNA libraries also confirm the high abundance of lipocalins. As previously described [9, 19], these proteins have important role in envenoming. The cuticle proteins identified can be related to the inflammatory response caused by macerated spicule proteins. The serpin protein may also be involved in the defense mechanism.

Besides the biochemical and pharmacological tests, the quality control of serum and vaccine production can be monitored by proteomic technologies [64], such as chromatographic analyses, bidimensional electrophoresis and immunoblotting, once they are able to detect protein degradation and also confirm the presence of specific antibodies. However, immunotherapy can be more effective if a better characterization of venom composition is performed, improving immunization procedures, increasing its specificity and reducing side effects. The new generation of high affinity antibodies against low abundant immunogenic toxins can be evaluated by an antivenomic approach [64, 65].

A novel approach is the investigation of post-translational modifications (PTM) that affect antigen recognition, given that many peptides presented to T cells by the major histocompability complex are post-translationally modified [66]. Glycosylation and phosphorylation are important PTMs of proteins, playing crucial roles in several biological processes, including cell recognition and signalling pathway [67, 68]. Some potential targets for cancer therapy are based on glycosylated and phosphorylated epitopes discoveries [59].

Otherwise, phosphorylated proteins are usually enriched by immunoprecipitation (mainly for phosphotyrosine peptides) or by chromatographic procedures, such as Strong Cation eXchange (SCX), Hydrophilic Interaction Liquid Chromatography (HILIC), Immobilized Metal Affinity Chromatography (IMAC) or Metal Oxide Affinity Chromatography (MOAC).

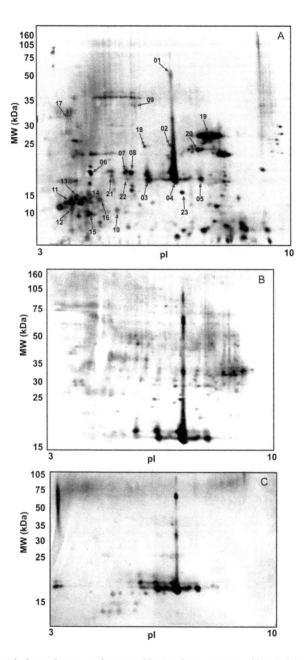

Figure 6. Bidimensional electrophoresis and immunoblotting from *Lonomia obliqua*'s bristle extract. (A): silver stained bidimensional gel (100 µg of protein applied) under non-reducing condition. Panels (B) and (C): PVDF immunoblotted 2D gels incubated with anti-Lonomic horse serum diluted 1:500 (B) or with anti-Lopap rabbit serum diluted 1:250 (C) [9].

Different metals may be used (iron, zirconium, gallium, etc) and peptides eluted by acidic or basic conditions, releasing mono-phosphorylated and multi-phosphorylated peptides, respectively [67].

The simultaneous screening of thousands of proteins from complex samples in a fast and sensitive manner can be performed using protein arrays. Amongst the different protein microarray applications are biomarker discovery, protein interaction studies, enzyme-substrate profiling, immunological profiling and vaccine development. As our interest is in the immune response, an antibody microarray can be used for identification of antigens that react specifically with the antibodies spotted on a solid support, with the complex formed then detected by fluorescence [59].

A large number of not yet identified proteins are considered as unknowns, but higher probabilities of identifications are reached when different methodologies are applied for analysis of complex samples. The combination of several proteomic techniques described here could improve the detection of immunogenic compounds and create new perspectives for effective immunotherapies.

3. *Lonomia obliqua* toxins

The *L. obliqua* caterpillar venom contain non-protein and protein components [21]. The procoagulant proteins present in the venom cause hemostatic disturbances mainly mediated by thrombin formation, the key enzyme of blood coagulation. Physiologically, thrombin generation from prothrombin occurs by assembling of the prothrombinase complex, which consists of factor Xa (catalytic factor), factor Va (non-enzymatic cofactor), calcium and a phospholipid membrane surface. Despite figured as a cascade of subsequent activation of coagulation factors, blood coagulation is currently conceived in a cell based-model [69]. In Figure 7 a simplified cascade model of hemostasis is illustrated, showing the known interactions of the *L. obliqua* venom and toxins.

Several molecules and activities were reported in bristles or hemolymph (Table 1). Some of them are related to the pathophysiology of the envenoming others to the development process of the animal such as regulation of the cell cycle [16]. Donato and colleagues [70] identified in the bristle extract a direct factor X activator which is calcium-independent, and a prothrombin activator. The prothrombin and factor X activators were later isolated and named Lopap and Losac, respectively [22, 24]. Interestingly, both molecules are no longer similar with any well-known procoagulant molecule from human or any other species.

3.1. Lopap: Functional characterization, recombinant production and bioinformatics analysis

Lopap (*Lonomia obliqua* Prothrombin Activator Protease) was purified from the bristle extract as a 69-kDa protein through gel filtration followed by reverse-phase chromatography. The purified protein was subjected to trypsin hydrolysis and partial amino acid sequences of N-terminal and internal fragments were obtained through Edman degradation [24].

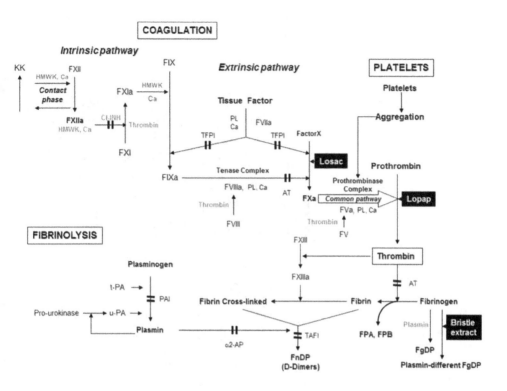

Figure 7. Schematic overview of hemostasis. Dark double-bars indicate where inhibitors act. HMWK = high molecular-weight-kininogen. PK = Prekallikrein. KK = Kallikrein. CI-INH = CI-inhibitor. TFPI = tissue factor pathway inhibitor. PL = phospholipids. Ca = calcium ions. AT = antithrombin. FDA = Fibrinopeptide A. FDB = Fibrinopeptide B. TAFI = Thrombin-activatable fibrinolysis inhibitor. FnDP = Fibrin degradation products. FgDP = Fibrinogen degradation products. a_2-AP = a_2-antiplasmin. Known interactions of the *L. obliqua* venom are indicated in the black boxes. Losac = *L. obliqua* Stuart factor activator. Lopap = *L. obliqua* prothrombin activator protease.

The recombinant protein (rLopap) was obtained in enzymatically active form as monomer of 21 kDa with a polyhistidine tag after purification by immobilized metal-chelate affinity chromatography. Partial amino acid sequences of native Lopap lead to identification of its respective clone from the cDNA library of *L. obliqua* bristles [20], encoding for a signal peptide (16 aa residues) and the mature protein (185 aa residues). cDNA of mature protein, consisting in a transcript with 603 bp open reading frame, was subcloned into the pAE vector and expressed in the bacteria E. coli BL21(DE3) with a fusion tag (His6). Protein was recovered in inclusion bodies after cell lysis and subjected to refolding and purification after solubilization in urea [71].

Interestingly, the deduced amino acid sequence of Lopap showed no similarity with other prothrombin activators or serine proteases, but was similar to lipocalin family members, either from insects or mammals [71]. Lopap sequence alignment with other lipocalins is shown in Figure 8. Members of lipocalin family usually share only about 30% of similarity in amino acid sequence, despite showing conserved secondary and tertiary structures. Furthermore, these proteins have in primary structure three characteristic conserved motifs [72].

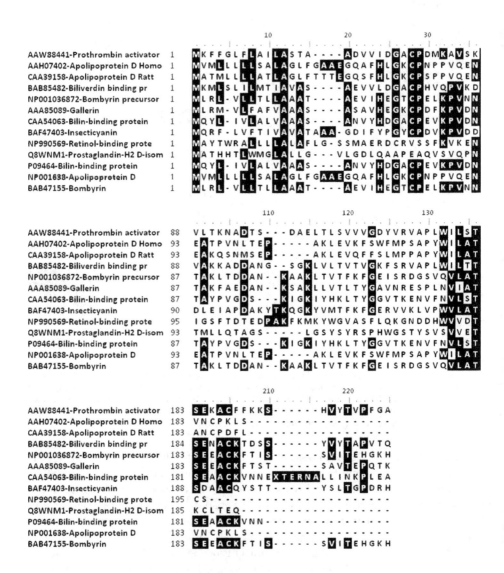

Figure 8. Amino acid sequence alignment of Lopap and other lipocalins. Sequences were accessed from protein data bank at NCBI and aligned using BioEdit [80].

Lopap's tridimensional structure obtained by molecular modeling has the characteristic fold of lipocalins, consisting in an eight stranded antiparallel β-barrel (Figure 9) with a hydrophobic pocket for binding of hydrophobic ligands. A serine protease catalytic triad was also predicted

[19]. Lopap is the first lipocalin described that displays proteolytic activity. On the other hand, through a peptide mapping approach based on lipocalin conserved motifs found in the Lopap's primary structure, a synthetic peptide was obtained (Figure 10), which has been proposed as a sequence signature among lipocalins, sharing a common role in cell protection and development process [73]. Other lipocalins that have been described with antiapoptotic activity share similar sequences, which have similar conformations in their tridimensional structures [73].

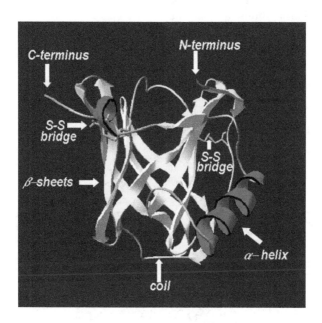

Figure 9. Model of the tridimensional structure of Lopap [81].

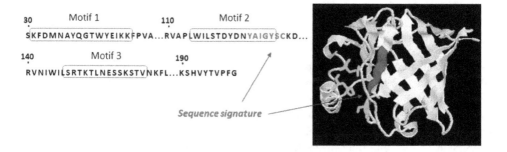

Figure 10. Lipocalin sequence signature highlighted among lipocalin conserved motifs identified in Lopap sequence [19] and in the model of tridimensional structure. Residues predicted in the catalytic triad are shown in green [25].

Lopap shows specific proteolytic activity towards prothrombin. It displays serine protease-like activity and activates human prothrombin through hydrolysis of Arg[284]-Thr[285] and Arg[320]-Ile[321] peptide bounds, generating active thrombin, without formation of the intermediate meizothrombin [24]. This mechanism is similar to prothrombin activation by FXa in absence of the prothrombinase complex (Figure 11), previously described [74]. This is the unique prothrombin activation mechanism described for an exogenous serine protease, which is independent of prothrombinase complex components. All other exogenous prothrombin activators (metalloproteases and serine proteases) currently described from snake venoms fit into four groups, sharing similar mechanisms of action [75].

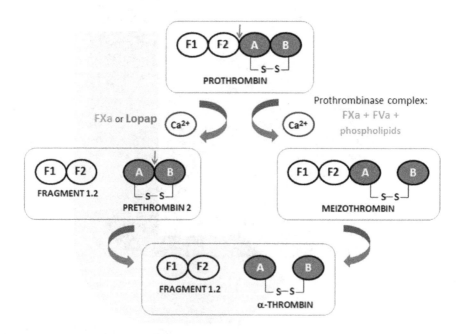

Figure 11. Prothrombin activation mechanisms indicating the Lopap hydrolysis sites and its generated products.

When administered *in vivo*, Lopap induces blood clotting into microvessels, resulting in fibrinogen consumption and blood incoagulability [76]. These effects resemble the consumption coagulopathy triggered by the whole venom, indicating the involvement of this pro-thrombin activator as a key toxin in envenoming [26]. In addition, Lopap is able to modulate endothelial cell responses promoting cell survival, IL-8, nitric oxide (NO) and PGI$_2$ release, expression of the cell adhesion molecules ICAM-1 and E-selectin, but not VCAM-1 and PCAM-1 [45, 77, 78]. Lopap also displays cytoprotective activity in neuthrophils and do not modify expressions of L-selectin and β2 integrin. Secretion of the proinflammatory cytokines IL-6 and TNF-α is not changed by Lopap treatment in both cell cultures [78]. Lopap seems to have no effect on modulation of coagulation and fibrinolysis through endothelial cell response, since it does not modify von Willebrand factor (vWF) and tissue plasminogen activator (t-PA) release or tissue factor procoagulant activity on endothelial cell surface [45, 77].

The Lopap-derived peptide obtained through chemical synthesis (Survicalin) reproduces the Lopap's modulation on endothelial cells and neutrophils cell, triggering antiapoptotic activity [73]. Survicalin also induces fibroblast responses, decreasing caspase-3 and increasing Bcl-2, Ki-67, IL-1β and the receptors for IL-8 and IL-6. Enhanced production of extracellular matrix proteins, such as collagen, fibronectin, tenascin and laminin is also induced by Survicalin in fibroblast culture [79].

3.2. Losac: Functional characterization

Losac is the first factor X activator purified from a lepidopter secretion. It was obtained from caterpillar's bristle extract as a single polypeptide chain protein of about 45 kDa [22]. Some years later, from a cDNA library of *L. obliqua* bristle transcripts [19], the specific clone encoding for Losac was identified and the recombinant protein produced in bacteria system (for details, see section 3.3). Studies using the native or recombinant form of Losac (rLosac) revealed specificity toward factor X [20, 22, 28]. Moreover, Losac had no effect on fibrin or fibrinogen, indicating its specificity for blood coagulation activation, and it was recognized by the antilonomic serum produced in Butantan Institute. Thus, it is plausible that this protein participates in the consumption coagulopathy observed in patients.

Biochemical characterization of Losac has shown that, although its sequence did not show an equivalent among other factor X activators, Losac possess a similar mechanism of action than RVV-X, a factor X activator purified from Russell's viper venom *Daboia russelli* [28, 82, 83]. Like RVV-X, factor X activation by Losac can be accelerated in the presence of calcium and phospholipids, two important cofactor in the assembling of blood coagulation complexes [69]. In spite of this, Losac can activate factor X independently of these cofactors. Moreover, both activators require a stable conformation of factor X and the presence of the Gla-domain of factor X for an appropriated activity. Interestingly, the cleavage fragments of factor X generated by both activators were quite similar. Although there are strong functional similarities, the major difference is in the structure of both activators. Apparently, Losac activates factor X through a serine protease-like activity, while RVV-X has a typical metalloproteinase structure [20, 28, 84].

A model proposed by Morita [83] and crystallographic studies of RVV-X [85] support the hypothesis that it primarily recognizes the calcium-bound conformation of Gla-domain in factor X through an exosite formed by the light chains, followed by the catalytic conversion of factor X to factor Xa. Despite the structural differences between Losac and RVV-X, it remains possible that they share a similar mechanisms for recognition of factor X involving calcium ions, phospholipids and the Gla-domain of factor X followed by its proteolytic conversion to active factor X.

Besides its role in coagulation [22, 28], Losac is also capable of inducing proliferation and inhibiting endothelial cell death while stimulating the release of NO, a known molecule with antiapoptotic activity [86, 87], and t-PA, a component of fibrinolytic-pathway involved in matrix remodeling [88]. The authors suggest that the cell proliferation and cell viability activities elicited by Losac are probably related to the NO liberation [22], since NO was also described as an endothelial survival factor, inhibiting apoptosis [86, 87]. Moreover, it was also observed that the production/expression of some important molecules involved in inflammation and coagulation systems such as ICAM-1, PGI2, DAF, IL-8, vWF and tissue factor were not affected by Losac.

It has been show that hemolymph from some insects can increase cell longevity by inhibiting apoptosis [89, 90]. The increase of *Spodoptera frugiperda* Sf-9 cell growth in almost 3-fold was reported after supplementation with *L. obliqua* hemolymph [91]. This effect was attributed to the presence of three factors with different activities: a potential antiapoptotic factor, a growth-promoting factor, and an enzyme that hydrolyzes sucrose. Furthermore, an antiapoptotic protein of 51 kDa was purified from *L. obliqua* hemolymph [36]. This protein was able to prevent apoptosis in Sf-9 cell culture induced by nutrient deprivation and by Actinomycin D. Later reports [37, 38] described in the hemolymph a potent antiviral activity against human virus.

3.3. Molecular cloning and heterologous expression of Losac

The production of Losac in a recombinant form was important due to the disadvantages of purifying Losac from bristle extract: the use of many caterpillars to prepare the bristle extract and the low yield of native Losac (0.3%) [22]. Cloning and production scheme to obtain rLosac is shown in Figure 12 [28].

Nucleotide and deduced amino acid sequences were compared with data banks in order to identify similar genes and their products. The analysis revealed a high similarity with members of the immunoglobulin-like superfamily of cell adhesion molecules (IgCAMs), especially with neural CAMs (NCAMs) [28]. Members of this group have diverse functions but none was associated with proteolytic activities [94]. Multiple comparison of the deduced amino acid sequence revealed different degrees of identity with IgCAMs: 26% of identity with L1-NCAM from humans, 34% with the protein neuroglian from *Drosofila melanogaster*; and 47-76% with hemolins from lepidopters [95-98]. Although no structural data was reported for Losac, a tertiary structure model was built through homology modeling based on crystal structure of *Hyalophora cecropia* hemolin (HcHemolin, Protein Data Bank code 1BIH). Both proteins share 76% of sequence identity [28] and the same multi-domain structure (four Ig-like domains: D1 to D4) and conserved motifs as shown in Figures 13 and 14. Both structures are composed of β-strands, arranged in a globular shape resembling a horseshoe (Figure 14A), akin to hemolin [97], axonin [99], and the four N-terminal Ig domains of neurofascin [100]. Because Losac shares its main sequence features with hemolins, it can be perfectly classified as one of them. It was demonstrated that Losac activate factor X in a similar way than RVV-X [28]. Nevertheless, unlike Losac, no hemolins or cell adhesion molecules were associated with proteolytic activities.

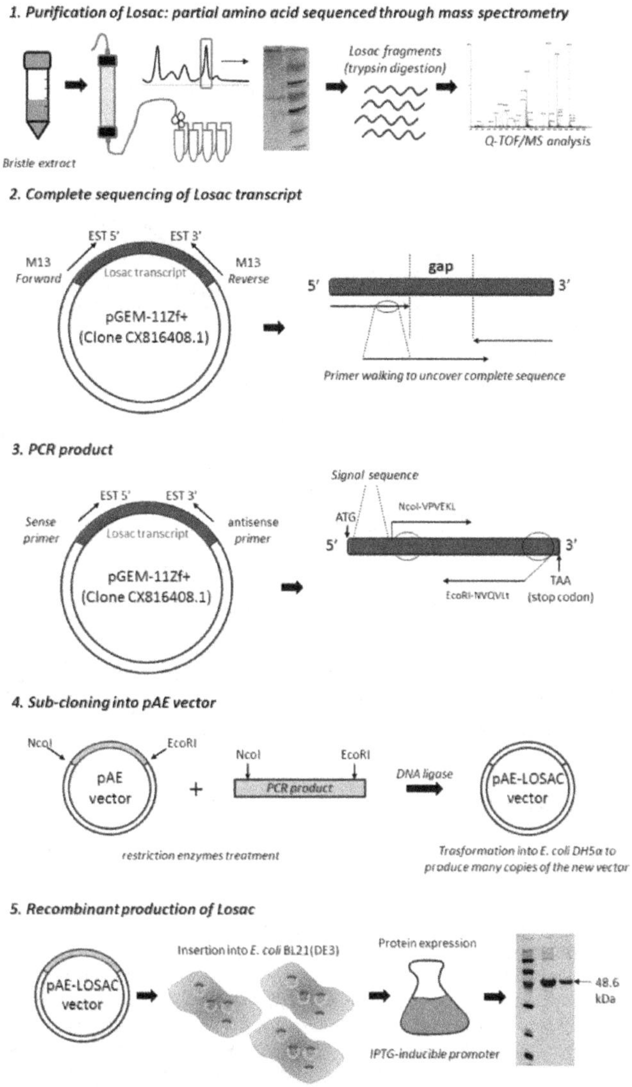

Figure 12. 1) After purification from bristle extract [22], Losac was submitted to a tryptic digestion to obtain partial amino acid sequences of internal fragments. Those sequences were obtained by mass spectrometry and used to screen the cDNA library to identify the transcript encoding the protein. The cDNA library was previously constructed with *L. obliqua* bristle mRNAs that were converted to cDNA and cloned into pGEM-11Zf+ plasmid [19, 20]. **2)** From this cDNA library, a transcript corresponding to the clone LOAH12B08 (GenBankTM accession number CX816408.1), matching the tryptic peptide sequences, was identified. However, this transcript was partially sequenced and the complete sequence was achieved through the *primer walking strategy* using a specific primer designed from an internal sequence of the transcript allowing uncover the complete sequence of Losac gene [92]. The nucleotide sequence has been deposited in GenBankTM (DQ479435), and the deduced protein has been deposited in the NCBI protein sequence database (ABF21073). **3)** The cDNA that encodes mature Losac was amplified by PCR using a sense and an

antisense primer designed according to the deduced N- and C-terminal sequences of the mature protein carrying *Nco*I or *Eco*RI restriction sites, respectively. **4)** The cDNA corresponding to mature Losac was sub-cloned into the pAE vector [93]. The PCR product and the pAE vector were restricted with *Nco*I and *Eco*RI, purified, and ligated with T4 DNA ligase and used to transform *E. coli* DH5α cells. **5)** The resulting pAE-Losac plasmid was used to transform *E. coli* BL21(DE3) cells. The recombinant protein was expressed in fusion to a minimal N-terminal His6-tag as a 48.6 kDa protein: Protein was recovered in inclusion bodies after cell lysis and subjected to refolding and purification after solubilization in urea.

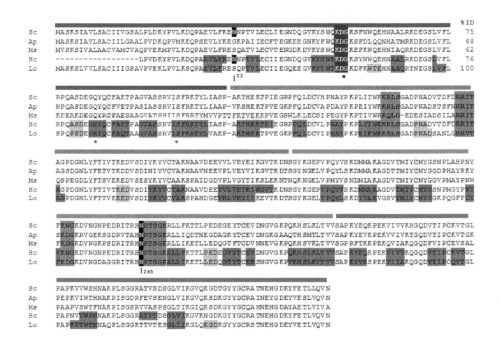

Figure 13. Multiple amino acid sequence alignment of Losac with other hemolin proteins. Sequence identity is shown for all proteins related to Losac. The structures encompass four constant-type immunoglobulin domains (cIg) (D1-D4). The bars above the sequences correspond to domains 1 (D1, blue), 2 (D2, green), 3 (D3, orange) and 4 (D4, red). Sequence identity is shown for all hemolins related to Losac (%ID). The α-helices and β-sheets observed in *Hyalophora cecropia* hemolin (Protein Data Bank code 1BIH) and Losac are shaded in light and dark gray, respectively. Previously it was predicted that hemolin contain conserved regions and motifs (28, 36). Conserved motifs are shown inside boxes according to the domain they belong. The eight cysteines (in yellow) form four intra-chain disulfide bridges. Losac conserves the well-known motifs involved in cell-adhesion mechanism (KDG motif in D1 and D3), as well as the highly conserved N-glycosylation site in D3 (Asn[265], in black). An additional N-glycosylation site (Asn[22]) is found in *H. cecropia* and *S. c. ricini* hemolins (D1). The LPS-binding site (NRTS motif: Asn[265], Arg[266], Thr[267] and Ser[268]) in D3 (orange). In D2 are located the KRLS cAMP/cGMP-dependent protein kinase phosphorylation site and the RRIT motif (green boxes). Positions of the residues forming the putative catalytic site are evidenced with an asterisk. Hemolin sequences (with abbreviation and GenBank™ accession numbers in brackets) are from: *Samia cynthia ricini* (Sc, BAE07175), *Antheraea pernyi* (Ap, AAS99343), *Manduca sexta* (Ms, AAC46915), Losac (Lo, ABF21073) and *H. cecropia* (Hc, AAB34817).

At this stage we can only speculate about the mechanism of action of Losac. One possibility is to evaluate structural features that might contribute to the Losac-induced factor X activation. Thus, a search for serine protease active-site was undertaken based on the *Catalytic Site Atlas* (CSA) program analysis. Three possible catalytic residues, Asp(D)[40], His(H)[72] and Ser(S)[90], were

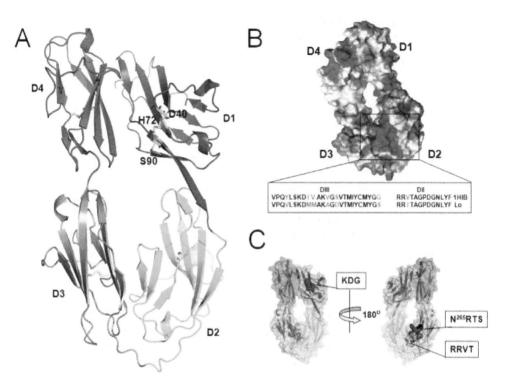

Figure 14. Three-dimensional structure model of Losac. **(A)** Cartoon view of the predicted model of Losac protein built from the structural coordinates of *H. cecropia* hemolin (PDB code 1BIH) using Modeller 9v1 [102]. Each domain (D1-D4) comprises α-helices and 7 strands arranged in 2 antiparallel β-sheets that are linked together by a disulfide bridge (shown in yellow stick). The residues of the putative catalytic triad predicted in D1 by the program *Catalytic Site Atlas* (http://www.ebi.ac.uk/thornton-srv/databases/CSA/) program, are indicated as stick cyan color by the one letter code followed by the residue number, D^{40}, H^{72}, S^{90} **(B)** Electrostatic potential surface of Losac showing inside the box the conserved phosphate-binding site at the D2-D3 interface domains, exactly as was demonstrated for HcHemolin [97]. **(C)** Surface view of Losac model evidencing the conserved adhesive motif KDG (deep blue) and the LPS-binding motif RRVT (green) and NRTS (chocolate), which contains the glycosylation site in N^{265}. All the figures were produced in Pymol v1.5 (http://www.pymol.org/).

located on D1 (Figure 13 and 14A) and could fit to such activity. Theoretical analysis (Figure 13 and 14) and observed results - total inhibition by PMSF as described in [28] - seem to suggest the presence of a serine protease-like active-site in Losac which would be responsible for the proteolytic activation of factor X [20, 28, 84]. Some molecular techniques, such as Site-directed mutagenesis [101], could be applied in an effort to understand the structural requirements for ligand binding and selectivity and identification of active site residues.

Functionally, hemolins were first associated with the insect immune system because of their over-expression after bacterial infection [103]. Due to their adhesion properties, some hemolins have been involved in the cell adhesion mechanisms. In the last two decades independent studies demonstrated that hemolins are multifunctional molecules involved in a diverse range of cell interaction [104-110]. The high identity among Losac and hemolins suggests that Losac could also assume some of these functions in *L. obliqua*. The adhesive properties of Losac probably are relevant to understanding the human umbilical vein endothelial cell responses observed in previous studies.

4. Biomedical applications

Studies on *L. obliqua* toxins with a molecular approach have applications beyond the pathophysiology and therapeutic perspectives of envenoming. As procoagulant proteins, Lopap and Losac can be useful as tools for developing clotting assays and diagnostic kits. Exogenous factor X activators, such as recombinant Losac, has also the potential to be used for detection of factor X deficiency and lupus anticoagulant [111]. In the case of Lopap, an exogenous prothrombin activator, two patents were applied to use this compound in diagnostic kits for detection of dysprothrombinemias using the native form purified from the venom as well as the recombinant form produced in bacteria. This prothrombin activator has also the potential to be used in clotting time assays, prothrombin assays, and to monitor patients anticoagulated with hirudin. A recent study suggests that exogenous procoagulant proteins could also be considered for therapeutic use to manage bleeding complications caused by anticoagulation therapy. Treatment with Lopap was able to reduce the bleeding time of rabbits anticoagulated with low molecular weight heparin, through direct prothrombin activation, bypassing factor Xa inhibition [112]. Patent information about those applications can be consulted in Table 3.

Modulation of cell responses triggered by *Lonomia* toxins can have valuable therapeutic and biotechnological applications. Promoting cell survival can be useful to improve cell culture technologies and vaccine productions, and for treatment of degenerative diseases. In addition, the effects of Lopap on extracellular matrix remodeling can be valuable to develop wound healing formulations and to regeneration issues (Table 3). For this approaches, design and synthesis of short peptides derived from Lopap amino acid sequence is an interesting task to minimize toxic and side effects and for production of this molecules for proofs of concepts, pre-clinical and clinical tests (Table 3). Isolating specific domains and sequences can also help to understand the multifunctional properties of the studied proteins and direct structure-function insights.

Unveiling the mechanisms of action and structure-function relationship of these multifunctional molecules may pointing out these molecules as promising candidates to development of new therapeutic drugs, reagents in diagnostic kits for coagulation dysfunctions, and biotechnological applications.

5. Concluding remarks

Nature has been finding ways to gift living beings with functions that are advantageous, regarding natural selection, mainly by evolutionary process. Among all the lepidopterans of

WIPO patent application	Publication date	Patent	Institutions involved
WO/2003/070746	08.28.2003	Purification and characterization of a prothrombin activator from the bristle of Lonomia obliqua: to be used in diagnosis kits for detecting plasma prothrombin in hemorrhagic state patients	Instituto Butantan (Brazil); Fapesp (Brazil) and Biolab Sanus Farmacêutica Ltda (Brazil)
WO/2006/021062	02.03.2006	Process for obtaining the recombinant prothrombin activating protease (rLopap) in monomeric form; the recombinant prothrombin activating protease (rLopap) as well as its amino acid sequence; the use of this protease as a defibrinogenase agent and the diagnosis kit for dysprothrombinemias	Instituto Butantan (Brazil); Fapesp (Brazil) and Biolab Sanus Farmacêutica Ltda (Brazil)
WO/2007/028223	03.15.2007	Lopap-based pharmaceutical compositions and uses thereof: it refers to the use of Lopap as modulators of cell death and degeneration caused by wounds, aging and external agents	Instituto Butantan (Brazil); Fapesp (Brazil) and Biolab Sanus Farmacêutica Ltda (Brazil)
WO/2009093189	07.30.2009	Peptides, compositions, and uses thereof: it refers to the uses of Lopap-derived peptides for regenerating tissues and wound repair	Instituto Butantan (Brazil); Fapesp (Brazil) and Biolab Sanus Farmacêutica Ltda (Brazil)

Table 3. International patents associated to Lopap and peptides derived from its amino acid sequence. Information was obtained from World Intellectual Property Organization (WIPO).

medical interest in the world, *Lonomia* sp. caterpillars (family: Saturniidae) is the only genus that causes dramatic damages in human blood coagulation [16, 113]. This feature is reflected in the diversity of toxins produced by the caterpillar and their unusual enzymatic properties.

Application of molecular approaches in the study of *L. obliqua* toxins has been a valuable strategy in understanding the biological means of these molecules for the source organism itself and the dynamic pathways in envenoming syndrome. On the other hand, this approach reveals these toxins as interesting tools for therapeutic and biotechnological applications. The best examples are Lopap (a prothromin activator lipocalin) and Losac (the only hemolin with proteolytic activity). If, in one hand, the molecular basis of target recognition and proteolysis of factor X and prothrombin by Losac and Lopap, respectively, needs to be further investigated, on the other hand, efforts need to be focused on understanding the pro-survival activity of both molecules.

Integrating transcriptomic, proteomic and microarray analysis will provide a wealth of valuable information about venom composition. Molecular cloning and expression of recombinant toxins from *L. obliqua* opens new perspectives in the identification and characterization of macromolecular fine structure of toxins and its implications for toxic activity as well as new action mechanisms and target cell binding that should be an area of rapid development. The next several years will likely see some very significant advances in this field and, in the future, those approaches will permit the identification of molecular mechanisms at a new level.

Acknowledgements

The authors thank the Brazilian founding agencies FAPESP and CNPq, INCTTOX program. L.C.C.-C. had a Post Doctoral fellowship from CAT-CEPID/FAPESP.

Author details

Ana Marisa Chudzinski-Tavassi, Miryam Paola Alvarez-Flores,
Linda Christian Carrijo-Carvalho and Maria Esther Ricci-Silva

*Address all correspondence to: amchudzinski@butantan.gov.br; miryam_paolaa@hotmail.com

Laboratory of Biochemistry and Biophysics, Butantan Institute, São Paulo, Brazil

References

[1] Diaz, J.H., The evolving global epidemiology, syndromic classification, management, and prevention of caterpillar envenoming. American Journal of Tropical Medicine and Hygiene, 2005. 72(3): p. 347-357.

[2] Hossler, E.W., Caterpillars and moths: Part II. Dermatologic manifestations of encounters with Lepidoptera. J Am Acad Dermatol, 2010. 62(1): p. 13-28; quiz 29-30.

[3] Balit, C.R., et al., Prospective study of definite caterpillar exposures. Toxicon, 2003. 42(6): p. 657-662.

[4] Hossler, E.W., Caterpillars and moths. Dermatologic Therapy, 2009. 22(4): p. 353-366.

[5] Ministério-da-Saúde, Acidentes por Lepidópteros, in Manual de diagnóstico e tratamento de acidentes por animais peçonhentos, FUNASA, Editor 2001, Fundação Nacional de Saúde (FUNASA): Brasil. p. 120.

[6] Zannin, M., et al., Blood coagulation and fibrinolytic factors in 105 patients with hemorrhagic syndrome caused by accidental contact with Lonomia obiqua caterpillar in Santa Catarina, southern Brazil. Thromb Haemost, 2003. 89(2): p. 355-64.

[7] Gamborgi, G.P., E.B. Metcalf, and E.J. Barros, Acute renal failure provoked by toxin from caterpillars of the species Lonomia obliqua. Toxicon, 2006. 47(1): p. 68-74.

[8] Goncalves, L.R., et al., Efficacy of serum therapy on the treatment of rats experimentally envenomed by bristle extract of the caterpillar Lonomia obliqua: comparison with epsilon-aminocaproic acid therapy. Toxicon, 2007. 50(3): p. 349-56.

[9] Ricci-Silva, M.E., et al., Immunochemical and proteomic technologies as tools for unravelling toxins involved in envenoming by accidental contact with Lonomia obliqua caterpillars. Toxicon, 2008. 51(6): p. 1017-28.

[10] Kowacs, P.A., et al., Fatal intracerebral hemorrhage secondary to Lonomia obliqua caterpillar envenoming - Case report. Arquivos De Neuro-Psiquiatria, 2006. 64(4): p. 1030-1032.

[11] Duarte, A.C., et al., Insuficiência renal aguda por acidentes com lagartas. Journal Brasileiro de Nefrologia, 1990. 12: p. 3.

[12] Kelen, E.M.A., Z.P. Picarelli, and A.C. Duarte, Hemorrhagic syndrome induced by contact with caterpillars of the genus Lonomia (saturniidae, hemileucinae). Journal of Toxicology-Toxin Reviews, 1995. 14(3): p. 283-308.

[13] Duarte, A.C., et al., Intracerebral haemorrhage after contact with Lonomia caterpillars. Lancet, 1996. 348(9033): p. 1033-1033.

[14] Carrijo-Carvalho, L.C. and A.M. Chudzinski-Tavassi, The venom of the Lonomia caterpillar: an overview. Toxicon, 2007. 49(6): p. 741-57.

[15] Fan, H.W. and A.C. Duarte, Acidentes por Lonomia, in Animais peçonhentos no Brasil: biologia, clínica e terapêutica dos acidentes, J.L. Cardoso, et al., Editors. 2003, Sarvier: Sao Paulo. p. 224-232.

[16] Alvarez Flores, M.P., M. Zannin, and A.M. Chudzinski-Tavassi, New insight into the mechanism of Lonomia obliqua envenoming: toxin involvement and molecular approach. Pathophysiol Haemost Thromb, 2010. 37(1): p. 1-16.

[17] Chudzinski-Tavassi, A.M. and L.C. Carrijo-Carvalho, Biochemical and biological properties of Lonomia obliqua bristle extract. Journal of Venomous Animals and Toxins Including Tropical Diseases, 2006. 12(2): p. 156-171.

[18] Reis, C.V., et al., In vivo characterization of Lopap, a prothrombin activator serine protease from the Lonomia obliqua caterpillar venom. Thromb Res, 2001. 102(5): p. 437-43.

[19] Reis, C.V., et al., Lopap, a prothrombin activator from Lonomia obliqua belonging to the lipocalin family: recombinant production, biochemical characterization and structure-function insights. Biochem J, 2006. 398(2): p. 295-302.

[20] Chudzinski-Tavassi, A.M. and M.P. Alvarez Flores, Exploring new molecules and activities from Lonomia obliqua caterpillars. Pathophysiol Haemost Thromb, 2005. 34(4-5): p. 228-33.

[21] Donato, J.L., et al., Lonomia obliqua caterpillar spicules trigger human blood coagulation via activation of factor X and prothrombin. Thrombosis and Haemostasis, 1998. 79(3): p. 539-542.

[22] Alvarez Flores, M.P., et al., Losac, a factor X activator from Lonomia obliqua bristle extract: its role in the pathophysiological mechanisms and cell survival. Biochem Biophys Res Commun, 2006. 343(4): p. 1216 23.

[23] Prezoto, B.C., et al., Antithrombotic effect of Lonomia obliqua caterpillar bristle extract on experimental venous thrombosis. Braz J Med Biol Res, 2002. 35(6): p. 703-12.

[24] Reis, C.V., et al., A prothrombin activator serine protease from the Lonomia obliqua caterpillar venom (Lopap) biochemical characterization. Thromb Res, 2001. 102(5): p. 427-36.

[25] Carrijo-Carvalho, L.C., Clonagem e expressão em levedura Pichia pastoris, obtenção de um peptídeo sintético, análise estrutural e avaliação de suas potenciais aplicações. PhD Thesis, in Interunidades em Biotecnologia2009, Universidade de São Paulo.

[26] Reis, C.V., et al., A Ca++ activated serine protease (LOPAP) could be responsible for the haemorrhagic syndrome caused by the caterpillar Lonomia obliqua. L obliqua Prothrombin Activator Protease. Lancet, 1999. 353(9168): p. 1942.

[27] Lilla, S., et al., Purification and initial characterization of a novel protein with factor Xa activity from Lonomia obliqua caterpillar spicules. J Mass Spectrom, 2005. 40(3): p. 405-12.

[28] Alvarez-Flores, M.P., et al., Losac, the first hemolin that exhibits procogulant activity through selective factor X proteolytic activation. J Biol Chem, 2011. 286(9): p. 6918-28.

[29] Seibert, C.S., E.M. Shinohara, and I.S. Sano-Martins, In vitro hemolytic activity of Lonomia obliqua caterpillar bristle extract on human and Wistar rat erythrocytes. Toxicon, 2003. 41(7): p. 831-9.

[30] Seibert, C.S., et al., Intravascular hemolysis induced by Lonomia obliqua caterpillar bristle extract: an experimental model of envenomation in rats. Toxicon, 2004. 44(7): p. 793-9.

[31] Seibert, C.S., et al., Purification of a phospholipase A2 from Lonomia obliqua caterpillar bristle extract. Biochem Biophys Res Commun, 2006. 342(4): p. 1027-33.

[32] Fritzen, M., et al., Lonomia obliqua venom action on fibrinolytic system. Thromb Res, 2003. 112(1-2): p. 105-10.

[33] Veiga, A.B., A.F. Pinto, and J.A. Guimaraes, Fibrinogenolytic and procoagulant activities in the hemorrhagic syndrome caused by Lonomia obliqua caterpillars. Thromb Res, 2003. 111(1-2): p. 95-101.

[34] Pinto, A.F., et al., Lonofibrase, a novel alpha-fibrinogenase from Lonomia obliqua caterpillars. Thromb Res, 2004. 113(2): p. 147-54.

[35] Gouveia, A.I.D., et al., Identification and partial characterisation of hyaluronidases in Lonomia obliqua venom. Toxicon, 2005. 45(4): p. 403-410.

[36] Souza, A.P.B., et al., Purification and characterization of an anti-apoptotic protein isolated from Lonomia obliqua hemolymph. Biotechnology Progress, 2005. 21(1): p. 99-105.

[37] Greco, K.N., et al., Antiviral activity of the hemolymph of Lonomia obliqua (Lepidoptera: Saturniidae). Antiviral Research, 2009. 84(1): p. 84-90.

[38] Carmo, A.C., et al., Expression of an antiviral protein from Lonomia obliqua hemolymph in baculovirus/insect cell system. Antiviral Res, 2012. 94(2): p. 126-30.

[39] de Castro Bastos, L., et al., Nociceptive and edematogenic responses elicited by a crude bristle extract of Lonomia obliqua caterpillars. Toxicon, 2004. 43(3): p. 273-278.

[40] Bohrer, C.B., et al., Kallikrein-kinin system activation by Lonomia obliqua caterpillar bristles: Involvement in edema and hypotension responses to envenomation. Toxicon, 2007. 49(5): p. 663-669.

[41] Berger, M., et al., Lonomia obliqua venomous secretion induces human platelet adhesion and aggregation. Journal of Thrombosis and Thrombolysis, 2010. 30(3): p. 300-310.

[42] Berger, M., et al., Lonomia obliqua caterpillar envenomation causes platelet hypoaggregation and blood incoagulability in rats. Toxicon, 2010. 55(1): p. 33-44.

[43] Veiga, A.B., et al., A catalog for the transcripts from the venomous structures of the caterpillar Lonomia obliqua: identification of the proteins potentially involved in the coagulation disorder and hemorrhagic syndrome. Gene, 2005. 355: p. 11-27.

[44] Pinto, A.F., et al., Novel perspectives on the pathogenesis of Lonomia obliqua caterpillar envenomation based on assessment of host response by gene expression analysis. Toxicon, 2008. 51(6): p. 1119-28.

[45] Fritzen, M., et al., A prothrombin activator (Lopap) modulating inflammation, coagulation and cell survival mechanisms. Biochem Biophys Res Commun, 2005. 333(2): p. 517-23.

[46] Crick, F., Central dogma of molecular biology. Nature, 1970. 227(5258): p. 561-&.

[47] Morozova, O., M. Hirst, and M.A. Marra, Applications of New Sequencing Technologies for Transcriptome Analysis, in Annual Review of Genomics and Human Genetics2009. p. 135-151.

[48] Marra, M.A., L. Hillier, and R.H. Waterston, Expressed sequence tags--ESTablishing bridges between genomes. Trends Genet, 1998. 14(1): p. 4-7.

[49] Adams, M.D., et al., Complementary DNA sequencing: expressed sequence tags and human genome project. Science, 1991. 252(5013): p. 1651-6.

[50] Junqueira-de-Azevedo Ide, L. and P.L. Ho, A survey of gene expression and diversity in the venom glands of the pitviper snake Bothrops insularis through the generation of expressed sequence tags (ESTs). Gene, 2002. 299(1-2): p. 279-91.

[51] Yang, Y., et al., EST analysis of gene expression in the tentacle of Cyanea capillata. FEBS Lett, 2003. 538(1-3): p. 183-91.

[52] Kozlov, S., et al., A novel strategy for the identification of toxinlike structures in spider venom. Proteins, 2005. 59(1): p. 131-40.

[53] Faria, F., et al., Gene expression in the salivary complexes from Haementeria depressa leech through the generation of expressed sequence tags. Gene, 2005. 349: p. 173-85.

[54] Burke, J., D. Davison, and W. Hide, D2 cluster: A validated method for clustering EST and full-length cDNA sequences. Genome Research, 1999. 9(11): p. 1135-1142.

[55] Miller, R.T., et al., A comprehensive approach to clustering of expressed human gene sequence: The sequence tag alignment and consensus knowledge base. Genome Research, 1999. 9(11): p. 1143-1155.

[56] Vera, J.C., et al., Rapid transcriptome characterization for a nonmodel organism using 454 pyrosequencing. Molecular Ecology, 2008. 17(7): p. 1636-1647.

[57] Schena, M., et al., Quantitative monitoring of gene-expression patterns with a complementary-DNA microarray. Science, 1995. 270(5235): p. 467-470.

[58] DeRisi, J.L., V.R. Iyer, and P.O. Brown, Exploring the metabolic and genetic control of gene expression on a genomic scale. Science, 1997. 278(5338): p. 680-686.

[59] Tjalsma, H., R.M. Schaeps, and D.W. Swinkels, Immunoproteomics: From biomarker discovery to diagnostic applications. Proteomics Clin Appl, 2008. 2(2): p. 167-80.

[60] Görg, A., et al., Two-dimensional electrophoresis with immobilized pH gradients for proteome analysis: A Laboratory Manual, 2002, Amersham Biosciences.

[61] Rabilloud, T., Two-dimensional gel electrophoresis in proteomics: old, old fashioned, but it still climbs up the mountains. Proteomics, 2002. 2(1): p. 3-10.

[62] Towbin, H., T. Staehelin, and J. Gordon, Electrophoretic transfer of proteins from polyacrylamide gels to nitrocellulose sheets: procedure and some applications. Proc Natl Acad Sci U S A, 1979. 76(9): p. 4350-4.

[63] Dass, C., Principle and Practice of Biological Mass Spectrometry2000, New York: Wiley.

[64] Gutierrez, J.M., et al., Snake venomics and antivenomics: Proteomic tools in the design and control of antivenoms for the treatment of snakebite envenoming. J Proteomics, 2009. 72(2): p. 165-82.

[65] Calvete, J.J., P. Juarez, and L. Sanz, Snake venomics. Strategy and applications. J Mass Spectrom, 2007. 42(11): p. 1405-14.

[66] Petersen, J., A.W. Purcell, and J. Rossjohn, Post-translationally modified T cell epitopes: immune recognition and immunotherapy. J Mol Med (Berl), 2009. 87(11): p. 1045-51.

[67] Eyrich, B., A. Sickmann, and R.P. Zahedi, Catch me if you can: mass spectrometry-based phosphoproteomics and quantification strategies. Proteomics, 2011. 11(4): p. 554-70.

[68] Lazar, I.M., et al., Recent advances in the MS analysis of glycoproteins: Theoretical considerations. Electrophoresis, 2011. 32(1): p. 3-13.

[69] Hoffman, M. and D.M. Monroe, Coagulation 2006: A modern view of hemostasis. Hematology-Oncology Clinics of North America, 2007. 21(1): p. 1-+.

[70] Donato, J.L., et al., Lonomia obliqua caterpillar spicules trigger human blood coagulation via activation of factor X and prothrombin. Thromb Haemost, 1998. 79(3): p. 539-42.

[71] Reis, C.V., et al., Lopap, a prothrombin activator from Lonomia obliqua belonging to the lipocalin family: recombinant production, biochemical characterization and structure-function insights. Biochemical Journal, 2006. 398: p. 295-302.

[72] Flower, D.R., The lipocalin protein family: Structure and function. Biochemical Journal, 1996. 318: p. 1-14.

[73] Chudzinski-Tavassi, A.M., et al., A lipocalin sequence signature modulates cell survival. FEBS Lett, 2010. 584(13): p. 2896-900.

[74] Krishnaswamy, S., et al., Activation of human prothrombin by human prothrombinase - influence of factor Va on the reaction mechanism. Journal of Biological Chemistry, 1987. 262(7): p. 3291-3299.

[75] Kini, R.M., The intriguing world of prothrombin activators from snake venom. Toxicon, 2005. 45(8): p. 1133-1145.

[76] Reis, C.V., et al., In vivo characterization of Lopap, a prothrombin activator serine protease from the Lonomia obliqua caterpillar venom. Thrombosis Research, 2001. 102(5): p. 437-443.

[77] Chudzinski-Tavassi, A.M., et al., Effects of lopap on human endothelial cells and platelets. Haemostasis, 2001. 31(3-6): p. 257-65.

[78] Waismam, K., et al., Lopap: a non-inflammatory and cytoprotective molecule in neutrophils and endothelial cells. Toxicon, 2009. 53(6): p. 652-9.

[79] Carrijo-Carvalho, L.C., et al., A lipocalin-derived Peptide modulating fibroblasts and extracellular matrix proteins. Journal of Toxicology doi:10.1155/2012/325250, 2012.

[80] Hall, T.A., BioEdit: a user-friendly biological sequence alignment editor and analysis program for Windows 95/98/NT. Nucl Acids Symp Ser, 1999. 41: p. 95-98.

[81] Reis, C.V., Clonagem, sequenciamento, expressão e caracterização parcial da estrutura do Lopap, um ativador de protrombina da lagarta Lonomia obliqua. PhD, in Biologia Molecular2002, Universidade federal de Sao Paulo.

[82] Kisiel, W., M.A. Hermodson, and E.W. Davie, Factor-X activating enzyme from Russells viper venom - Isolation and characterization. Biochemistry, 1976. 15(22): p. 4901-4905.

[83] Morita, T., Proteases which activate factor X, in Enzymes from snake venoms, G.S. Bailey, Editor 1998, Alaken Inc.: Fort Collins, Colorado. p. 179-208.

[84] Tans, G. and J. Rosing, Snake venom activators of factor X: an overview. Haemostasis, 2001. 31(3-6): p. 225-33.

[85] Takeda, S., T. Igarashi, and H. Mori, Crystal structure of RVV-X: An example of evolutionary gain of specificity by ADAM proteinases. Febs Letters, 2007. 581(30): p. 5859-5864.

[86] Dimmeler, S., et al., Upregulation of superoxide dismutase and nitric oxide synthase mediates the apoptosis-suppressive effects of shear stress on endothelial cells. Arterioscler Thromb Vasc Biol, 1999. 19(3): p. 656-64.

[87] Rossig, L., et al., Nitric oxide inhibits caspase-3 by S-nitrosation in vivo. J Biol Chem, 1999. 274(11): p. 6823-6.

[88] Bobik, A. and V. Tkachuk, Metalloproteinases and plasminogen activators in vessel remodeling. Curr Hypertens Rep, 2003. 5(6): p. 466-72.

[89] Rhee, W.J., E.J. Kim, and T.H. Park, Kinetic effect of silkworm hemolymph on the delayed host cell death in an insect cell-baculovirus system. Biotechnol Prog, 1999. 15(6): p. 1028-32.

[90] Kim, E.J., H.J. Park, and T.H. Park, Inhibition of apoptosis by recombinant 30K protein originating from silkworm hemolymph. Biochem Biophys Res Commun, 2003. 308(3): p. 523-8.

[91] Maranga, L., et al., Enhancement of Sf-9 cell growth and longevity through supplementation of culture medium with hemolymph. Biotechnol Prog, 2003. 19(1): p. 58-63.

[92] Strauss, E.C., et al., Specific-primer-directed DNA sequencing. Analytical Biochemistry, 1986. 154(1): p. 353-360.

[93] Ramos, C.R.R., et al., r-Sm14 - pRSETA efficacy in experimental animals. Memorias Do Instituto Oswaldo Cruz, 2001. 96: p. 131-135.

[94] Shapiro, L., J. Love, and D.R. Colman, Adhesion molecules in the nervous system: Structural insights into function and diversity, in Annual Review of Neuroscience2007. p. 451-474.

[95] Haspel, J., et al., Critical and optimal Ig domains for promotion of neurite outgrowth by L1/Ng-CAM. J Neurobiol, 2000. 42(3): p. 287-302.

[96] Bieber, A.J., et al., Drosophila neuroglian: a member of the immunoglobulin superfamily with extensive homology to the vertebrate neural adhesion molecule L1. Cell, 1989. 59(3): p. 447-60.

[97] Su, X.D., et al., Crystal structure of hemolin: a horseshoe shape with implications for homophilic adhesion. Science, 1998. 281(5379): p. 991-5.

[98] Li, W., et al., Cloning, expression and phylogenetic analysis of Hemolin, from the Chinese oak silkmoth, Antheraea pernyi. Dev Comp Immunol, 2005. 29(10): p. 853-64.

[99] Freigang, J., et al., The crystal structure of the ligand binding module of axonin-1/ TAG-1 suggests a zipper mechanism for neural cell adhesion. Cell, 2000. 101(4): p. 425-33.

[100] Liu, H., P.J. Focia, and X. He, Homophilic adhesion mechanism of neurofascin, a member of the L1 family of neural cell adhesion molecules. J Biol Chem, 2011. 286(1): p. 797-805.

[101] Sarkar, G. and S.S. Sommer, Double-stranded DNA segments can efficiently prime the amplification of human genomic DNA. Nucleic Acids Res, 1992. 20(18): p. 4937-8.

[102] Sali, A., Comparative protein modeling by satisfaction of spatial restraints. Mol Med Today, 1995. 1(6): p. 270-7.

[103] Faye, I. and M.R. Kanost, Function and regulation of hemolin, in Molecular Mechanism of immune Responses in insect, P.T.a.H. Brey, D., Editor 1998, Chapman and Hall: New York. p. 173-188.

[104] Ladendorff, N.E. and M.R. Kanost, Bacteria-induced protein P4 (hemolin) from Manduca sexta: a member of the immunoglobulin superfamily which can inhibit hemocyte aggregation. Arch Insect Biochem Physiol, 1991. 18(4): p. 285-300.

[105] Daffre, S. and I. Faye, Lipopolysaccharide interaction with hemolin, an insect member of the Ig-superfamily. FEBS Lett, 1997. 408(2): p. 127-30.

[106] Yu, X.Q. and M.R. Kanost, Binding of hemolin to bacterial lipopolysaccharide and lipoteichoic acid. An immunoglobulin superfamily member from insects as a pattern-recognition receptor. Eur J Biochem, 2002. 269(7): p. 1827-34.

[107] Lanz-Mendoza, H., et al., Regulation of the insect immune response: the effect of hemolin on cellular immune mechanisms. Cell Immunol, 1996. 169(1): p. 47-54.

[108] Bettencourt, R., et al., Cell adhesion properties of hemolin, an insect immune protein in the Ig superfamily. Eur J Biochem, 1997. 250(3): p. 630-7.

[109] Schmidt, O., et al., Role of adhesion in arthropod immune recognition. Annu Rev Entomol, 2010. 55: p. 485-504.

[110] Kanost, M.R., et al., Isolation and characterization of a hemocyte aggregation inhibitor from hemolymph of Manduca sexta larvae. Arch Insect Biochem Physiol, 1994. 27(2): p. 123-36.

[111] Chudzinski-Tavassi, A.M., et al., Exogenous factors affecting hemostasis: therapeutic perspectives and biotechnological approaches, in Animal toxins: State of the Art. Pesrpectives in Health and Biotachnology, M.E. De Lima, et al., Editors. 2009, UFMG: Belo Horizonte. p. 495-523.

[112] Andrade, S.A., et al., Reversal of the anticoagulant and anti-hemostatic effect of low molecular weight heparin by direct prothrombin activation. Brazilian Journal of Medical and Biological Research, 2012.

[113] Zannin, M., et al., Blood coagulation and fibrinolytic factors in 105 patients with hemorrhagic syndrome caused by accidental contact with Lonomia obliqua caterpillar in Santa Catarina, Southern Brazil. Thrombosis and Haemostasis, 2003. 89(2): p. 355-364.

Chapter 6

Venom Bradykinin-Related Peptides (BRPs) and Its Multiple Biological Roles

Claudiana Lameu, Márcia Neiva and
Mirian A. F. Hayashi

Additional information is available at the end of the chapter

1. Introduction

The kallikrein-kinin system is an extensively studied biological pathway and involves a multi-protein complex, which includes serine proteinases from tissue and plasma. These proteinases act on substrates as kininogens (high and low molecular weight), releasing the active kinins. The main kinin is the nonapeptide bradykinin (BK).

Several studies aiming to evaluate the biological activities of the kinins revealed that this peptide is implicated in diverse physiological processes as regulation of blood pressure, cardiac, and renal function. Due to its ability to increase the vascular permeability by acting on endothelial cells, BK is correlated to several pathological processes including inflammation. These actions have been observed and described in both mammals and rodents [1].

The knowledge on the role of BK in various biological pathways as coagulation cascade, blood pression regulation, and central nervous system modulation and signaling has been significantly improved, leading to the identification of BK receptors and posterior development of drugs targeting its pathways [2].

This research was mainly driven by scientific studies on animal venoms, which lead to the identification of the BK-related peptides (BRPs). The best, and maybe also the first, example of such contribution was the discovery of the bradykinin-potentiating peptides (BPPs), first described in *Bothrops jararaca* venom [3, 4]. The BPPs are proline-rich oligopeptides that inhibit the angiotensin-converting enzyme (ACE), and that are responsible for the hypotensive effect of the *Bothrops* genus snake venoms. The pharmacological effects of these peptides have been studied since 70's [3, 4], and allowed not only to the discovery of the neuropeptide BK [5], but also to the development of the first active site-directed inhibitor of ACE as drug for the treatment of human hypertension [6]. In fact, several other drugs derived from venom toxins, with or without modifications, are also commercially available (*e.g.* Captopril, Ancrod, and Prialt) [7]. Moreover, the study of toxins has widely contributed to the identification of new targets with therapeutic potential in mammals, as well as it has allowed to the understanding and discovery of the biochemical pathways involving these targets.

Since then, the BPPs/BRPs have been found in several snake venoms, and also in wasps and frogs, by using either biochemical or/and recombinant DNA techniques [8-11]. For instance, molecular cloning studies using cDNA libraries of four species of snakes from Crotalinae family showed evidences that these bioactive peptides are expressed by orthologous genes [12]. The cloning of orthologous precursors from different snakes from *Bothrops* and *Crotalus* genus allowed the identification of several new BPPs sequences [13-15], and some of them was shown to display different specificity toward each active sites of the somatic ACE ectoenzyme [16]. This was believed to be a great opportunity for the development of a new generation of antihypertensive drugs.

The employment of recombinant DNA techniques were also fundamental to first determine the structure of the precursor protein of BPPs, which was found to contain several sequences of BPPs distributed as tandem repeats, followed by a C-type natriuretic peptide (CNP) at the C-terminus of this precursor molecule [15]. In contrast to other members of the natriuretic peptide (NP) family, CNP is synthesized in the brain and has hypotensive effect with no significant diuretic or natriuretic actions [17]. Moreover, Northern blot analysis of several snake tissues demonstrated the presence of similar BPPs-CNP precursor mRNA in non-venomous tissues, such as the central nervous system (CNS) [14]. In *situ* hybridization studies also detected the presence of the BPP/CNP-precursor mRNA in regions of snake brain correlated with neuroendocrine functions, such as the ventromedial hypothalamus, paraven-tricular nuclei, paraventricular organ, and subcommissural organ [14]. Analogous CNP precursor mRNAs was also described in similar regions in rat and human brains [18].

These studies suggesting the potential expression of BPPs in snake CNS stimulated us to investigate the putative target(s) of these peptides. Based on the *in vivo* biodistribution studies showing the preferential accumulation of BPPs in the rat kidney, and also a significant presence in the brain, the first studies were conducted in theses tissues leading to the description of several completely new potential targets and pathways, as the nicotinic acetylcoline receptors [19], L-argininosuccinate synthase [20], and an orphan G protein-coupled receptor (GPCR) [20]. The importance of both NO release for the antihypertensive effects of BPPs [20-22], and also the involvement of the GPCRs, namely B2 receptor and M1 muscarinic receptor (mACh-M1), in vasodialtion were demonstrated [23].

Together all these data collected during the last decade showed the pharmacologial signifi-cance of the BPPs and, more importantly, that the biological effects of these peptides, although first believed, are not limited to the inhibiton of the somatic ACE [2]. The high variability of molecular structures of these peptides reflecting in different specifities is an indicative that there are still more to be discovered regarding the biological effects of this peptides family.

In this book chapter, we intend to gather the most important results obtained up to now, thanks to the isolation and characterization of BPPs from diverse organisms and to the knowledge accumulated, while searching for new targets for these molecules.

2. The discovery of the snake venom BK and BRPs

The main function of snake venoms is still believed to be the immobilization of preys to ensure feeding. The snake venoms are composed of a complex mixture of proteins and biologically active peptides [24, 25]. The study of the pathophysiological mechanisms of poisoning and molecular characterization of toxins from the venom of *Bothrops jararaca* resulted in many scientific contributions of great importance, and among them, stand out the discovery of BK [5] and the discovery of the first BRPs, more specifically the BPPs produced by the snake venom glands, [4, 26] whose synergistic action is capable of causing a sharp drop in blood pressure of small animals, for instance mammalian preys.

The BPPs are molecules able to enhance some pharmacological activities of BK, as the action of contractile smooth muscle of guinea pig ileum evaluated in *ex vivo* assays [26], and also *in vivo*, acting in the CNS, cardiovascular, and antinociceptive systems [27, 28]. The isolation of the first BPPs expressed by the *Bothrops jararaca* venom glands was described in the early 60's, and they were initially coined as Bradykinin Potentiating Factors (BPF) due to their ability to potentiate the effects of BK ignoring at that time the fact that these molecules were composed by amino acid residues [26]. Only in early 70's, when their primary sequence were determined, which allowed to characterize them as peptide molecules, they were re-named as BPPs [3]. Since then, several peptides presenting similar structural characteristics have been identified from the venom of these snakes and also from other snakes belonging to several different genus [12, 13, 29-31]. Interestingly, they had also been described in wasps and frogs [8-11]. Typically, the BPPs are peptides of 5-14 amino acid residues [32]. In general, all known naturally occuring BPPs could be classified into two groups: (i) peptides of small molecular size like BPP-5a from the venom of *Bothrops jararaca*, whose structural characteristic is a pyroglutamic acid at the N-terminal and a proline residues at the C-terminal of molecule, and (ii) peptides consisting of about ten amino acid residues, with a pyroglutamic acid at the N-terminal and a notable high content of proline residues [32], which gives to them some resistance to hydrolysis by amino-peptidases, carboxypeptidases, and also endopeptidases [33].

2.1. cDNA cloning, identification and characterization of BRPs

The BK and its related peptides, *e.g.* the BRPs, are widely found in venomous animals, for instance in snakes, lizards, frogs, and insects [10, 13, 34]. In general, they include several sequences, either showing only one single amino acid substitution compared to BK or, in some cases, presenting just a frugal sequence similarity, but with unquestionable biological/ functional correlation, for instance, acting on the same pathway or even same target protein. In fact, these sequence variations were verified either by *de novo* sequencing of several BRPs found in snake venoms [32] or by analysis of the deduced amino acid sequences of cDNAs cloned from venomous glands [12, 14, 15], and in some cases by using both strategies [30, 34].

The pharmacological evaluations revealed that even acting in the same pathway, they can show distinct biological activities compared to BK, including potentiating its effects by inhibition of its degradation or by acting on receptors and/or molecules involved in the BK signaling pathway, including activating or blocking the BK receptors [10, 35]. As such, the BRPs also include BPPs and the Bradykinin Inhibitor Peptides (BIPs) [13, 29, 32, 36].

2.2. BPPs and BIPs

Helokinestatins are a family of proline-rich peptides (PRPs) found originally in the lizard venom (*Heloderma suspectum*) that display the function of inhibiting the BK actions on the vascular smooth muscle [35]. Synthetic replicates of all helokinestatins were found to antagonize the relaxation effect observed following BK application to a rat arterial smooth muscle preparation, and hence, represent a family of BRPs also known as BIPs [34].

In contrast, BPPs firstly described and isolated from the venom of the Brazilian snake *Bothrops jararaca* are mainly known due to their ability to potentiate the biological effects of BK [3, 26]. These BRPs are one of the most outstanding group of PRPs, as they were used as structural and functional template/model for the development of a drug employed up to now for the treatment of human hypertension [6].

Although functionally related to the BK and also present with the NPs in the same precursor protein, the helokinestatins are quite different from the snake venom BPPs [12, 30, 31, 37-39]. PRPs with the same BK inhibitory characteristics have also been described in the 'venomous' secretion of two species of anguid lizards, the *Texas alligator* (*Gerrhonotus infernalis*) and the *Giant Hispanolian galliwasp* (*Celestus warreni*) [38]. Although the primary structural variation of the peptides from these species, they share several common features [34]. For instance, they are peptides rich in prolyl residues (30-50%), which confer rigidity and order to the spatial structure features, and also a measure of resistance to generalized proteolysis. They all possess a Pro-Arg dipeptide motif at the C-terminus, which is quite different from the C-terminal Ile/Val-Pro-Pro motif present in most BPPs C-terminus extremity [12]. The high degree of conservation of these structural core features across phylogeny suggest a fundamental biological function for this group of peptides in the lizards venoms. Among the two closely-related species of helodermatid lizards, several helokinestatins have fully-conserved primary structure, while several others present different sequences. Similarly to the BRPs from amphibian skin [40] and snakes venom [13, 14, 30], helokinestatins compose tandem repeat domains in their respective precursor proteins, probably reflecting discrete exons within the genomic DNA. As already mentioned, some tandem repeats are composed by identical primary structure, while some others exhibit significant amino acid substitutions. prominent And this process of exon multiplication might facilitate the molecular diversity, by permitting the expression of site-mutated isoforms, which is a phenomenon often described for bioactive peptide-encoding genes, as also observed for the glucagon gene in vertebrates [41].

Cloning and alignment of cDNAs encoding BRPs precursors from the venom gland and brain of a pit viper have allowed observing a higher degree of sequence consevation for the regions not including the bioative peptides, and a higher variation in the primary structure of these biological active peptides [14]. These results were shown to be in good agreement with the accelerated evolution hypothesis suggested by Ohno and colleagues [42]. According to this hypothesis, the more frequent occurrence of nucleotide nonsynonymous substitutions in the coding regions compared to the untranslated regions (UTRs) of the genes allows specific genes to evolve in an accelerated fashion to attain unique physiological activity. On the other hand, despite the consequent changes in the BRPs sequences observed, both the high content of proline residues and the biological activities correlated to BK effects are still maintained (Figure 1). Another highly conserved region involves the sequence of the NPs always present in C-terminus extremity of all known BRPs precursor proteins [12-15, 30, 31, 34, 36].

2.3. BRPs and NPs

The NP system consists of three types of hormones [atrial NP (ANP), brain or B-type NP (BNP), and C-type NP (CNP)], and three types of receptors [NP receptor (R)-A, NPR-B, and NPR-C]. Both ANP and BNP are circulating hormones secreted from the heart, whereas CNP is basically a neuropeptide. The NP system plays pivotal roles in cardiovascular and body fluid homeo-stasis. The ANP is secreted in response to an increase in blood volume, and acts on various organs to decrease both water and Na^+, resulting in restoration of blood volume. The family of NPs were originally Na^+-extruding hormones in fishes; however, they evolved to be volume-depleting hormones promoting the excretion of both Na^+ and water in tetrapods, in which both are always regulated in the same direction. Vertebrates expanded their habitats from fresh water to the sea or to land during evolution. The structure and function of osmoregulatory hormones have also undergone evolution during this ecological evolution [43].

Members of the NPs family have been detected in several snake venoms, and they have been shown to be located in the same precursor protein containing multiple BRPs sequences. While this organization was demonstrated for many species of viperid snakes, including members of the genera *Bothrops, Crotalus, Lachesis, Agkistrodon,* and *Trimeresurus* [12, 13, 15, 30, 31], it may extend to some other taxa such as *Bitis gabonica,* that was also shown to have BPPs in their venom [12, 13, 15, 30, 31, 44]. The presence of NPs in some elapid snakes venom (*Dendroaspis*) has also been described [45]. It has been shown that the venom-derived NP precursors from helodermatid lizard have a structural organization similar to that found in many BRPs precursors from viperid snake venoms. However, the additionally encoded tandem-repeat peptides are non-canonical BPPs, based on their primary structural characteristics or in terms of the amino acid cleavage site, presenting characteristics of a recognition site typical of propeptide convertase enzymes, that eventually might be the potential responsible for the release of the mature BIPs from the respective biosynthetic precursors [36]. However, the BPP/CNP biosynthetic precursors of the bushmaster (*Lachesis muta*), the tropical rattlesnake (*Crotalus durissus terrificus*), and the massasauga rattlesnake from desert (*Sistrurus catenatus ewardsi*) showed that, in addition to the classical BPPs and a NP sequence, they all also encode single copies of a BIP exhibiting a closer structural similarity, and a propeptide convertase cleavage site that allows the release of the BIP helokinestatins, whose sequences are [TPPAGPDVGPR] or [TPPAGPDGGPR] [36] (Figure 1). On the other hand, putative heloki-nestatins peptides could not be identified in the BPP/CNP precursor of snakes as *Bothrops jararaca, Bothrops jararacussu,* and *Agkistrodon blomhoffi* (Figure 1). The phylogenetic analysis

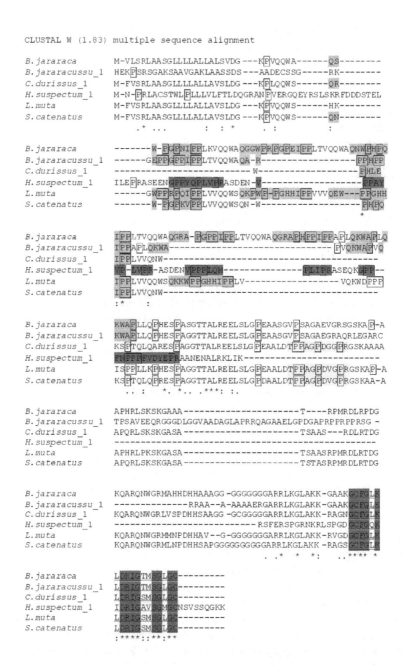

Figure 1. Alignment of the amino acid sequences of BRPs precursors. Organization of the BPP/CNP and helokines-tatin/CNP precursor from venoms, indicating the mature BPPs (grey), helokinestatins, BIPs (red), and CNP (underlined) Note that precursors of *C. durissus*, *L. muta*, and *S. catenatus* present both mature BPPs and BIPs in their sequences. The conserved amino acid seqences compared to fragments involving the CNP region are highlighted (pink) and the proline residues are indicated by boxes.

presented here separates NPs precursor of different species into three distinct groups, those which contain in their precursor sequence (i) only BPPs, (ii) only BIPs, and/or (iii) both BRPs, *i.e.* BPPs and BIPs (Figure 1 and 2), suggesting that mutations in the coding regions of BRPs were important for the adaptative changes along evolution of the venom system [40].

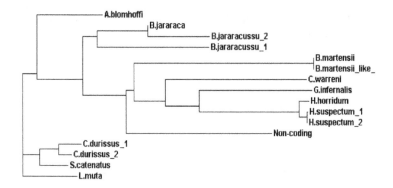

Figure 2. Phylogenetic tree based on BRP/NPs precursors. The phylogenetic analysis was performed using T-Coffee - Multiple Sequence Alignment avaiable at http://www.ebi.ac.uk/Tools/msa/tcoffee/. In this analysis, protein sequences of BRP/CNP precursors from reptiles were used. The NPs precursor of *A. blomhoffi. B. jararaca*, and *B. jararacussu* contain only BPP sequences; *C. warreni, G. infernalis, H. suspectum*, and *H. horridum* contain only BIP sequences and, *C. durrissus, L. mutta*, and *S. catenatus* contain both BRPs sequences, *i.e.* BPP and BIP. Despite the aligned *B. martensii* sequence was partial, and therefore does not contain the sequence coding for BRPs, this species was found to express the closest related precursor sequence to those containing only BIPs and to the *B. jararaca* non-coding RNA homologous to BPP/CNP precursor.

2.4. Molecular evolution of genes encoding BRPs

Here, we take BPP/CNP and helokinestatin/CNP precursors as examples to illustrate the evolution of a gene, since BPP/CNP precursor is also expressed in other tissues of *Bothrops jararaca* besides the venom gland, incluing brain and spleen [14, 15]. A high similarity was oberved for the BPP/CNP cDNAs isolated from brain and venom gland, although they are not identical to each other as it should be expected [14]. Three out of the five BPP isoforms present in the brain precursor (BPPs of 5, 10 and 13 amino acid residues) were identical to those found in the venom gland precursor [14]. Moreover, most of insertions/deletions and point mutations were observed within the BPP/CNP coding region, suggesting an effect of a Darwinian-type accelerated evolution frequently observed intra-specie [42, 46] This process has been widely observed, since a number of neuropeptides and hormones, such as the NPs [15, 47] and the vascular endothelium growth factor [48] evolved into toxins in the venom gland of poisonous animals.

It is believed that BRPs from the venom may be considered the toxin counterparts of endogenous peptides. It has also been suggested that both CNP and BPPs could be physiologically associated, to perform fluid homeostasis and regulation of the vascular tonus, since BPP/CNP precursor are present in regions of the snake brain showed to be involved in the control of these activities, as described for the mammalian CNS [14].

It is known that in the process of evolution, several mutations may occur in the genes, some of which not affecting the mature protein sequence, while others might lead even to the generation of messenger RNAs that are not translated into proteins. In 2000's it was first shown that non-coding RNAs can be involved in several roles including repression of genes, catalysis, regulation of the development process, among others [49]. A non-coding mRNA showing a sequence similarity to the BPPs precursor of the pit viper *Bothrops jararaca* was cloned by us from the venom gland [Genbank Acc. No. AY310916.1]. This long RNA sequence does not encode a protein, since several stop codons were observed for all possible reading frames (Figure 3).

```
G T R Stop K L P N N Stop S N S L I V L T I R T I N Q R N C L S S E F V G A S S L E A F E K R L D R H L S E
Met V F S N S I I I L V Y Stop L T P I E I L C K T Q L W T R L Q Stop R F Y L F Y L L I K F L C R S S F K T
R S G Stop Q H K Stop N N I S T F Q N Stop N Q S Stop K A Y N F Y Y I A F A P V S E G A Met S Y I A L P K L
T L S K S H K S H R R K N S E Q A Met L I Stop S A W R S E V N F V W K T L S L E A K Met A Met G I H L D Met
Y Met S R Y S I N Q S N K E I R K Y S F S N F P S I L N S H R C T S G E R S K F K N S D I Stop F V L R L Y F
Y L F F F F S N L S C C L Stop G I I S K E V L L A I L Q L F Y W V T P A V V E V G K K K I S V I Y F W N R E
D N A I Q W Met I V F S N N N Stop T A F Stop C N F L F S I L F F Stop F Q A Y L F S V R A H R V G L L R V P
S V R Q C H L A G P R A R A F S V A A P A L W N Q L P S E V H T A S S L L A F R K L L K T H L C R Q V W G V
Stop A A L C P W A F D C D C V W C E Stop N K L R S F Y T V N Stop Stop F Stop L I Y V Met F F I L V L L L
Stop T A L S L V R R A A Y K F N K Stop Stop Stop Stop Y S L T N V V V F Y K Q Stop E L H A Q S Stop V V K
Met Stop N V S Stop A D S W L G E F K F L L I S E K K F C F L E V F A G F Y C W H I Met Stop I P Q Stop S A
G D G Met Stop P F S K V L P F Stop C S T G H I N F P L D C R I S W Q K V I S R Q P D R T L Q Q F F Met V Y
S H G I S S L Q K V K K N N N S H K R A Stop H Stop K S E F F Stop N E A L S Y L N V L S L H G G Met E N P P
Met G I Q S L L P Y H R L G P K F S S A S W L P P Q F F D R R S S S S G R T V Stop L L F H V L Stop E N Y L
P R P V R G G R G F S I P P C R E R T S G K T T L L S P A W V D S E S T N G T Y P S P Q K S P W E G V Stop A
I S Y Stop A S T I C C R D L L P N E P S A A A N Stop S I L Stop F L T K E D G E A H V A A L Q I S L Q R R L
R A P S C R R R H L A A V F E L R P H P I R G H F W T S P A I H P G I P G P H S L S P V G T S T Stop R S H F
S V G E V L G R G W T E F C A P K S S Y L T Stop Y I P R C H I Y I D R Y K Y T Q H K K T R P P T T S S P R Q
R T V S G N G E N Q L R Met P S R L L F V H S C H A Stop V L C K S Y L Y I V Y K E I F I K K H Y P
N Stop T K R G R A L Stop T F F F S F F S S Stop I V G R K E K K I Met D D L I T C L A F Stop S D A V F L S
E Met G G L N K A F R Y H G Stop E Stop K Met H L Stop R N R T L S T K G V K E F D A P F L Met F L A E I K K
P L F K E V L R I K K K K K K K K K K K K K K K K K K K K K K K K K K
```

Figure 3. Potential deduced amino acid sequences from the all possible frames of the non-coding RNA homologous to the BPP/CNP precursor coding RNA both from *Bothrops jararaca*. Total mRNA obtained from *Bothrops jararaca* venom gland higher than 2 Kb length was used to construct the long cDNA library. The complete cDNA sequence of the BPP/CNP precursor (clone NM 96) was used as template to screen about 2 x 10^6 clones, allowing the identification of four independent positive clones containing identical inserts of approximately 3.5 Kb [15]. All potential frames of the *Bothrops jararaca* BPP/CNP-related pseudogene mRNA presents a high content of stop codons (bold), as shown by the representative amino acid deduced sequence from one of the possible reading frames. No signal peptide could observed, although several methionine (Met) residues (underlined) that does not seem to represent an initial of translation were found.

These non-coding RNAs are usually transcribed by a gene known as a pseudogene, which are often found in the genomes of several life forms, including bacteria, plants, insects, and vertebrates [50]. The pseudogene is a sequence that is present in the genome, and it is typically characterized by presenting high similarity with one or more functional gene paralogs. The pseudogene can be derived from gene duplication occurred by two different pathways: retrotransposition or duplication of genomic DNA [50].

The regulation of the expression of a functional gene showing sequence similarity to a pseudogene has been reported [51]. Generally, the sequence similarity between functional genes and pseudogenes is observed at the 5' UTR fragment. However, for the comparison of the non-coding RNA and the RNA coding for BPP/CNP precursor, a high similarity was observed only for the 3' UTR sequence (Figure 4). Moreover, the non-coding RNA is of approximately 3.5 Kb, while the RNA coding for the BPPs precursor is of about 1.8 Kb, and its RNA expression was found to be about 6-fold higher than that of the 3.5 Kb non-coding RNA.

Nevertheless, it is also possible that non-coding RNA of 3.5 Kb [GenBank Acc. No. AY310916.1] may also ensures the stability of the functional coding messenger RNA, since the stability of messenger RNAs is preferably controlled by factors present in the 3' UTR region [52].

```
non-coding   TATCTCTTACTAGGCCTCTACTATCTGCTGCAGGGACCTTCTACCAAACGAACCCTCAGC
BPP-coding   TGCTTT------GGCCT---------------GAAGCT--------CGA--CCGCATC
             *   *        ***** *            * * **         *** ** ** *

non-coding   TGCAGCGAATTGATCGATCTTGTAATTTCTAACGAAGGAGGACGGAGAGGCCCACGTGGC
BPP-coding   GGCACCA---TGAGCGGCCT-----------------GGGCTGCTGAAGCCGTCGAGG-
             *** *    *** ** **               ** *  ** *** ** **

non-coding   CGCTCTGCAAATTTCTCTCCAGCGGCGCCTGCGTGCGCCAAGCTGCCGAAGGCGACACCT
BPP-coding   --------------------GCGGCGC-----------------GAAGGCGACACCT
                                 * * * * *                 *************

non-coding   GGCGGCGGTCTTTGAACTGAGACCCCACCCCATCCGCGGACATTTCTGGACATCCCCTGC
BPP-coding   GGCGGCGGTCTTTGAACTGAGACCCCACCCCATCCGCGGACATTTCTGGACATCCCCTGC
             ************************************************************

non-coding   AATTCATCCAGGGATCCCAGGCCCACACAGCCTGTCTCCTGTTGGTACGAGCACTTGAAG
BPP-coding   AATTCATCCAGGGATCCCAGGCCCACACAGCCTGTCTCCTGTTGGTACGAGCACTTGAAG
             ************************************************************

non-coding   AAGCCATTTTTCAGTGGGAGAGGTTCTGGGTAGAGGTTGGACTGAGTTTTGCGCCCCCAA
BPP-coding   AAGCCATTTTTCAGTGGGAG-GGTTCTGGGTAG-GGTTGGACTGAGTTTTGCGCCCCCAA
             ******************** ************ ***************************

non-coding   ATCTTCCTATCTTACCTAATATATACCTAGATGCCATATATATATCGATAGATACAAGTA
BPP-coding   ATCTTCCTATCTTACCTAATATATACCTAGATGCCATATATAT ATCGATAGATACAAGTA
             *********************************************  *************

non-coding   CACCCAACACAAAAAAACACGCCCGCCCACAACTTCTTCCCCGCGACAAAGGACGGTGTC
BPP-coding   CACCCAACACAAAAAA-CACGCCCGCCCACAACTTCTTCCCCGCGACAAAGGACGGTGTC
             **************** *******************************************

non-coding   GGGAAACGGAGAGAACCAGCTGCGTATGCCATCGAGATTGCTGTTTGTACATTCGTGTCA
BPP-coding   GGGAGACGGAGAGAACCAGCTGCGTATGCCATCGAGATTGCTGTTTGTACATTCGTGTCA
             **** ******************************************************

non-coding   CGCATAAATGTATTTAAGTTGTAAAAGCTATTTATATATTGTTTATAAAGAGATATTTAT
BPP-coding   CGCATAAATGTATTTAAGTTGTAAAAGCTATTTATATATTGTTTATAAAGAGATATTTAT
             ************************************************************

non-coding   AAAAAAAAATTTTATTTATGTAAACTAAACGAAAAGGGGTCGAGCATTGTAAACTTTTTT
BPP-coding   AAAAAAAAATTTTATTTATGTAAACTAAACGAAAAGGGGTCGAGCATTGTAAACTTTTTT
             ************************************************************

non-coding   TTTTTCCCCTTTCTCATCCTGAATAGTCGGAAGGAAAGAAAAAAAAA-TCATGGACGATT
BPP-coding   TTT--CCCCTTTCTCATCCTGAATAGTCGGAAGGAAAGAAAAAAAAAAATCATGGACGATT
             ***  *************************************************  *****

non-coding   TGATAACCTGTCTTGCATTTTAATCGGATGCAGTCTTTCTTTCCGAAATGGGTGGTTTGA
BPP-coding   TGATAACCTGTCTTGCATTTTAATCGGATGCAGTCTTTCTTTCCGAAATGGGTGGTTTGA
             ************************************************************

non-coding   ACAAGGCTTTTAGATACCACGGATAAGAGTAAAAAATGCATTTGTAAAGAAATCGCACGC
BPP-coding   ACAAGGCTTTTAGATACCACGGATAAGAGTAAAAAATGCATTTGTAAAGAAATCGCACGC
             ************************************************************

non-coding   TATCTACCAAAGGTGTAAAAGAACCCGACGCGCCGTTTTTGATGTTTTTAGCAGAAATAA
BPP-coding   TATCTACCAAAGGTGTAAAAGAACCCGACGCGCCGTTTTTGACGTTTTTAGCAGAAATAA
             ****************************************** *  *************

non-coding   AAAAA-CCACTTTTCAAAGAGGTCCTCAGAATT
BPP-coding   AAAAAACCACTTTTCAAAGAGGTCCTCAGAATT
             ***** ***************************
```

Figure 4. Partial sequence alignment of the *Bothrops jararaca* **BPP/C NP-related pseudogene mRNA and mRNA coding for BPP/CNP precursor.** Alignment of the nucleotide sequences of a segment of the pseudogene mRNA (non-coding: upper sequence) and the mRNA coding for BPP/CNP precursor (BPP-coding: lower sequence) was performed using the Clustaw W program, available at http://www2.ebi.ac.uk/clustalw/. Identical nucleotides are indicated by "*" and insertions or deletions are represented by gaps (-).The boldface type letters indicate the region with higher similarity between the RNA sequences, corresponding to about 97% identity in this region.

In the BPP/CNP precursor of *Bothrops jararaca*, the pentapeptide BPP-5a [QKWAP] that was used as template for the development of the antihypertensive drug captopril, is found duplicated, *i.e.*, there are two copies of the same peptide in a single precursor protein. It is believed that this peptide might have a special importance in the venom of snakes belonging to the *Bothrops* genus, since it is also found repeated three times in isoform 1 [GenBank Acc. No. AY310914.1], and four times in isoform 2 [Genbank Acc. No. AY310915.1] of the precursors isolated from *Bothrops jararacussu* venom glands (Figure 5). In fact, BPP-5a is a potent potentiator of the BK effects in isolated guinea pig ileum, and also *in vivo* [29].

Figure 5. Partial nucleotide and amino acid sequences of the BPP/CNP of precursor from*Bothrops jararacussu* **(isoform 1 and 2).** Shaded in grey, the amino acid sequences of the C-type natriuretic peptide (CNP). Sequences of new putative BPPs are shown in green, and the underlined sequences correspond to other previously known BPPs. In red, the pentapeptide BPP-5a that was found in duplicate in the pit viper precursor and, triplicate and quadruplicate in the isoforms 1 and 2, respectively, of the precursor from *Bothrops jararacussu*. Symbol "M" represents the initial methionine and (-) the stop codon.

3. BRPs as structural model for drug development

The discovery of the potential inhibitory action of BPPs on ACE brought a great interest in these natural peptides, since the importance of ACE in blood pressure control and the urge to develop a therapy for cardiovascular disease, as hypertension, was iminent [2].

At that time, among the identified peptides were BPP-9a, under the generic name of teprotide, and the BPP-5a, which was also one of first BPP to be characterized. Assays using these peptides showed that BPP-9a was more effective and had a longer lasting effect in blood pressure compared to BPP-5a [53]. Therefore, BPP-9a was used in the first clinical demonstration of the potential use of BPPs for the hypertension control in humans, showing a significant antihypertensive effect [54, 55].

However, on that time it was demonstrated that the therapeutic utility of BPP-9a was limited by the lack of activity by oral administration and the high cost of its synthesis [54, 56, 57]. Therefore, the pharmaceutical development of a non-peptide inhibitor of ACE orally effective was essential. Thus, molecular structure of the BRPs, namely BPP-5a and BPP-9a, were studied by Cushman and Ondetti [3, 58], who suggested specific interaction of the proline, present at the C-terminal of these peptides, with the ACE active site [59]. Thus, captopril was synthetized by simple addition of a chelator radical to a dipeptide containing a proline residue (BPP carboxy-terminal amino acid) [59]. Unodoubtely captopril was a blockbuster drug that inspired the creation of generations of mimetic antihypertensive compounds [2].

4. Biological activities of BRPs

4.1. Interference of BRPs in the renin-angiotensin and kallikrein-kinin system

The ACE (EC 3.4.15.1) is mainly expressed in vascular endothelium in epithelial cells of the proximal tubules of the kidney, brain, and intestinal cells [60]. This enzyme is responsible for conversion of angiotensin I (Ang I) to angiotensin II (Ang II), and for the degradation of BK. Therefore, this enzyme has roles in both renin-angiotensin and kallikrein-kinin system [61].

The renin-angiotensin system (RAS) is composed by a set of peptides, enzymes, and receptors, that are involved in the control of the extracellular fluid and blood pressure [62]. The formation of the effector peptide of this system occurs initially by the action of the renin released by the kidneys [62] that acts on the angiotensinogen produced in the liver [63]. This leads to the generation of the decapeptide Ang I, which then is cleaved by ACE to form the octapeptide Ang II, a potent antihypertensive molecule [64]. Ang II actions is mediated by the angiotensin receptors AT1 and AT2. The binding of Ang II to the AT1 receptor triggers several cellular processes, among them vasoconstriction, protein synthesis, cell growth, regulation of renal function, and electrolyte balance [65]. Ang II also acts as a neurotransmitter and as a neuro-regulador, modulating the central control of the blood pressure, influencing the sympathetic activity, salt appetite, and thirst [65].

The kallikrein-kinin system (KKS) is a metabolic cascade in which the tissue and plasma kallikrein release vasoactive kinins from both high and low molecular weight kininogens. The nonapeptide BK, derived from the cleavage of the high molecular weight kininogen by kallikrein, is the major plasma kinin playing a role in the KKS [66].

Kinins are involved in various physiological and pathological processes, including vasodilation, increased vascular permeability, release of plasminogen activator of tissue type (t-PA), and nitric oxide (NO) and arachidonic acid metabolism, mainly due to their ability to activate endothelial cells [66]. Thus, the kinins participate in the physiologically regulation of blood pressure, cardiac and renal functions, and also in pathological processes as inflammation [66].

The several pharmacological activities of kinins are mediated basically by the their binding to two types of specific receptors (B1 and B2 receptors), prior to their fast metabolization by various peptidases [67].

Actions such as vasodilation and hypotension are mediated by the B2 receptor by releasing of NO, prostacyclin, and endothelium-derived hyperpolarizing factor (EDHF). On the other hand, the actions mediated by the B1 receptors include important roles in angiogenesis, inflammation, and septic shock [68]. Moreover, unlike B2 receptor, B1 receptor is not constitutively expressed, and its expression is induced by mediators of inflammation in conditions of injury [68].

The primarily responsible for the degradation of BK are the peptidases (zinc metallopeptidases) including ACE [67]. Since the early 90's, it is well known that somatic ACE has two active sites, the N-terminus (N-site) and the C-terminus (C-site) active sites [69]. Although *in vitro*, the two active sites are equally effective to convert Ang I to Ang II, as well as to degrade the BK into BK_{1-7} and BK_{1-5} [70], the N-site is several times more effective to hydrolyze other bioactive peptides, such as the AcSDKP, a negative regulatory factor for differentiation and proliferation of hematopoietic stem cells [71].

Thus, ACE inhibitors as BPPs inhibit not only the generation of Ang II, but also potentiate the effects of BK, by inhibiting its degradation. Therefore, the physiological effects of the angiotensin system are decreased (since there is no formation of Ang II), and the physiological effects of KKS are potentiated (due to inhibition of the BK degradation). In contrast, the BIPs, most known as helokinestatins, inhibit KKS by blocking the B2 receptor (Figure 6).

4.2. Mechanisms of action underlying the antihypertensive effect of BRPs

Although ACE inhibition is a relevant mechanism to explain the activity of most BPPs, and despite of their high primary sequence similarity [53, 72], as previously suggested, the BPPs show remarkable wide variety of mechanistic pathways that could explain the antihypertensive activity of BPPs at molecular level [13, 14, 19, 20, 22, 23, 29, 73-76]. Definitely the biological effects of BPPs and the consequent pharmacological importance of their activity are not limited to and it cannot be explained solely based on their ability to inhibit ACE [2].

The differences were first observed when comparing the selectivity of the BPPs encoded by the neuronal BPP/CNP precursor protein [*e.g.* BPP-5a, BPP-10c, BPP-11e, BPP-12b, and BPP-13a] [14] by the different active sites of the somatic ACE and the corresponding biological activity of these peptides evaluated by their ability to potentiate the contractile effect triggered by BK in isolated guinea pig ileum. For instance, the BPP-5a was shown to be much less effective ACE inhibitor compared to BPP-13a, although presenting one of the most potent potentiator effects of BK in *ex vivo* experiments. In contrast, BPP-10c is an excellent selective inhibitor of the C-terminal active site of somatic ACE, and its BK potentiating effect is very similar to that observed for both BPP-5a and BPP-12b, which were shown to be selective for

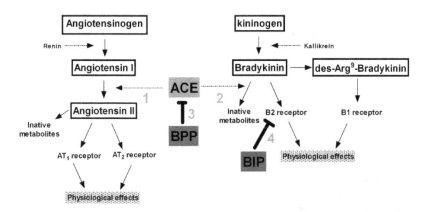

Figure 6. Schematic representation of ACE roles on the renin-angiotensin and kallikrein-kinin systems, and the potential sites for interference by BRPs (BPPs and BIPs). A) *Conversion of angiotensin I into angiotensin II, 2) BK degradation 3) ACE inhibition by BPPs. 4) B2 receptor antagonism by BIPs.* Physiological effects on the renin-angiotensin system mediated by AT1 receptors include vasoconstriction, sodium and water retention, release of aldosterone, increased sympathetic nerve activity, among others, while those mediated by AT2 receptors include cell differentiation, vasodilation, among others. The effects on the kallikrein-kinin system, mediated by kinins action on B2 receptor include vasodilation and hypotension via release of NO, prostacyclins and endothelium-derived hyperpolarizing factor (EDHF). Due to the ACE inhibition by BPPs, the physiological effects of angiotensin system are decreased (with no formation of angiotensin II) and the physiological effects of kallikrein-kinin are potentiated (by inhibition of BK degradation). In contrast, BIPs action on B2 receptor blocked BK effects. Adapted from [132-134].

the ACE N-terminal active site. In the same way, besides the weak BK potentiation effects of BPP-11e, it is also not among the best inhibitors of ACE, and no preference for any of the active sites of ACE was observed for this peptide [29].

Later on, in 2007, molecular studies of the antihypertensive activity of the BPPs, namely BPP-7a and BPP-10c, brought noteworthy information on the molecular mechanism underlying the action of these peptides at cellular level. In fact, these BPPs have a strong and sustained antihypertensive activity in awake spontaneously hypertensive rats (SHRs), but they do not prevent the formation of Ang II from Ang I *in vivo*, showing that they do not need to affect the physiological functions of ACE to promote the decrease of the blood pressure in these animals. Furthermore, for BPP-10c, we have also shown that the dose necessary to produce the antihypertensive effect is lower than that required to inhibit ACE *in vivo* [77], suggesting the participation of other putative targets determining this particular pharmacological effect.

This finding was reinforced by the studies conducted to clarify the biological distribution of BPP-10c using a I^{125} labeled analog, which showed that this peptide accumulated in various rat organs such as brain, liver, testis, and kidney, even after pre-saturation of the potential active sites of ACE with a specific inhibitor of this enzyme, namely captopril [78].

This stimulated us to conduct studies aiming to identify new potential molecular targets for snake BPPs. So, it was shown that at least three BPPs, namely BPP-10c, BPP-12b, and BPP-13a, are able to bind to the enzyme argininosuccinate synthase (AsS) modulating positively its activity [20, 75].

The AsS is the rate-limiting step enzyme responsible for providing the substrate for the nitric oxide synthase (NOS) that produces NO [79, 80]. Guerreiro and colleagues also demonstrated that blood pressure decrease promoted in SHRs by BPP-10c administration is due to the increased bioavailability of L-arginine required for the production of NO [20], which is a potent vasodilator agent [81]. Later it was demonstrated that other BPPs also induces NO production to determine the antihypertensive effect [21, 75].

Moreover, at least for the BPP-5a-induced NO production, the involvement of both B2 receptor and mACh-M1, without any involvement of AsS, was recently demonstrated [23]. BPP-13a induces NO production through a mechanism that involves activation of subtype M3 muscarinic receptor (mACh-M3), triggering the raise of the free intracellular calcium concentration ($[Ca^{2+}]_i$) that is able to activate NOS and to provide the substrate for NO production by modulating the AsS activity [75].

Both BPP-11e and BPP-12b do not stimulate NO production, but the $[Ca^{2+}]_i$ mobilization assays suggest that these peptides are agonists of a membrane receptor involved in the release of EDHF, and other functions involving the modulation of gene expression and activation of different NOS enzymes is expected [82]. As BPP-12b modulates positively AsS activity only at very high concentrations, this should not be its main mechanism of action [75].

Since the BPP-9a has ACE as main target for its biological actions, based on its potent inhibitory activity against this enzyme also showing selectivity for the C-terminal active site [73], we suggest that it is possible to suggest this pathway to explain the antihypertensive effect and BK potentiation of BPP-9a (teprotide). Moreover, it has no effect on the AsS induced intracellular calcium and it also does not interfere with NO production.

Apparently all BPPs share the ability to decrease arterial pressure [21, 56, 75, 77, 83], through the amplitude of the antihypertensive effect caused by BK, each related peptides is different. But, unfortunately, the mechanisms of action of other BPPs are still less understood up to now [75].

4.3. Peripheral and central biological activities of BPPs

Changes in mean arterial pressure (MAP) promoted by some BPPs are accompanied by a significant reduction in heart rate (HR) [23, 75, 77] rather than by an HR increase, as it would be expected by the response of the baroreceptors to the hypotension [84]. The fact is that *in bolus* injections of BPPs decrease both MAP and HR of awake SHRs, and BPPs expression in the same precursor protein of a brain expressed peptide as CNP suggests a CNS role for these peptides. In fact, recently it was shown that the BPP-10c is able to promote the release of the neurotransmitters GABA and glutamate, which are known to participate in the regulation of cardiac and vascular autonomic systems, leading to decline MAP and HR of SHRs [22]. According to Lameu and collaborators, BPP-10c-induced decrease of MAP results from this BRP-induced interference in the autonomic nervous system, provoking subsequent changes in HR and baroreflex control [22, 74].

Arterial baroreflex is one of the most important regulatory mechanisms in the cardiovascular system, mainly by triggering a coordinated sympathetic and parasympathetic tone response on the heart and vessels [85-90].

The CNS is connected to the heart through two different groups of nerves, the parasympathetic and sympathetic systems. Stimulation of parasympathetic nerves determines the decrease of HR, of contraction force of atrial muscle, and of conduction of impulses through the atrioventricular node, and at the same time, it also causes the increase of the time delay between the atrial and ventricular contraction, and the reduction of blood flow through the coronary arteries, which maintains the nutrition of the myocardium. All these effects can be summarized by saying that the parasympathetic stimulation decreases all the activities of the heart. On the other hand, the stimulation of sympathetic nerves has exactly the opposite effects on the heart, leading to an increased HR, increased contraction force, and increased blood flow through the blood vessels [87, 91].

It was observed that BPP-11e causes a slight reduction in MAP, but surprisingly with a strong reduction in HR [75], suggesting a BPPs action in specialized muscle cells located in the sinoatrial region (pacemaker) of the heart, which is a special region of the heart that controls the cardiac frequency [92]. Although the heart has its own intrinsic control systems, it can operate under neural influences, therefore effectiveness of the cardiac action can be significantly modified by regulatory pulses from the CNS [92]. Thus there is also possible that the BPP-11e has an effect on the stimulation of the parasympathetic system and/or in the decreasing of the sympathetic system stimulation, leading to a reduction of the HR and a slight decrease in MAP, observed after *in vivo* injection of this peptide [75].

A more detailed study of the BPPs effects on the CNS was performed for the BPP-10c, in which intracerebroventricular administration was shown to produce similar effects to those observed for higher doses injections of this peptide by intravenous route. In our interpretation, this data suggested the involvement of the CNS in the pathway underlying these biological effects [74].

Aiming to explain tbe BPPs effects on CNS, *Lameu et al.* have also conducted studies to demonstrate that the BPP-10c acts through activation of an unidentified $G_{i/o}$-coupled receptor present in neuronal cells, and that this effect was independent of both ACE inhibition and B2 receptor activation. Peptide–receptor binding resulted in the activation of calcium influx and release of intracellular calcium by calcium-induced calcium release (CICR) mechanism, which was shown to involve the activation of the ryanodine- or IP3-sensitive calcium stores and also the inhibition of adenylate cyclase [74]. However the specific target GPCR could not be identified yet.

On the other hand, affinity chromatography, using immobilized BPP-10c, associated with mass spectrometric and immunoblot analyses, allowed the identification of two important targets of BPP-10c, namely the AsS in the kidney cytosol [20] and the synapsin in the brain (Figure 7) [93]. AsS, together with argininosuccinate lyase (AsL), is part of the urea cycle in the liver and of the arginine-citrulline cycle, the major source of arginine in the renal cells and citrulline–NO cycle, which is the main source of NO in other cells, including endothelial and neuronal cells [94].

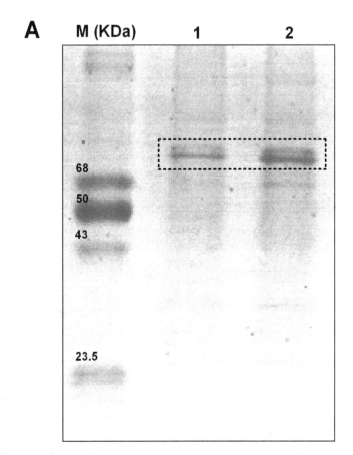

A   M (KDa)   1   2

B

Lane	Elution type	Protein	Molecular Mass	Swiss Protein Database	Covarage (%)
1	Competition	Synapsin	74.1	NP_062006	43
2	Lowering of pH	Synapsin	74.1	NP_062006	57

Figure 7. Synapsin binds to BPP-10c. (**A**) SDS-PAGE analysis of brain rat cytosolic proteins submitted to HiTrap-BPP-10c affinity chromatography. Rat brain cytosol preparation to affinity chromatography using the HiTrap NHS-activated HP resin to which BPP-10c was immobilized by chemical conjugation **M (KDa)**, molecular mass markers; **lane 1**, protein eluted by competition using 5 mg of BPP-10c; **lane 2**, protein eluted with 100 mM glycine, 0.5 M NaCl pH 3.0 (elution buffer: by lowering the pH). (**B**) protein identification by mass spectrometric analysis of the bands enclosed in the box in the panel **A**. The 74-kDa major protein that bounds to BPP-10c was identified as synapsin, by trypsin digestion and peptide mass fingerprint analysis.

AsS is a ubiquitously expressed enzyme, present in many tissues, including brain and kidney [94]. *In vivo* BPP-10c administration in SHRs animal models results in increase of plasma arginine level [20] and augmented NO production in brain tissues, as well as in neuronal and endothelial cells [20, 22].

NO is generated in the citrulline-NO cycle by NO synthase (NOS) using L-arginine as a substrate. Three isoforms of NOS have been described: Ca^{2+}-dependent endothelial (eNOS) and neuronal (nNOS) isoforms [95] and inducible NOS. The expression and activity of the latter are induced by inflammatory stimuli, independent of the cytosolic Ca^{2+} concentration [96].

NO is mainly involved in the regulation of local and systemic vascular resistance in sodium balance, and hence in blood pressure control [97], since it is one of the smooth muscle relaxing factors released by the endothelium, which diffuses to the adjacent smooth muscle cells promoting vasodilatation [98, 99].

Nevertheless, NO has been attributed to other various functions, including non-cholinergic and non-adrenergic smooth muscle relaxation, reduction of arterial pressure, and signal transmission in the CNS [100]. NO-mediated actions in the CNS include central vascular regulation [101] and baroreflex control of HR [102]. Antihypertensive activity, based on the facilitated release of both GABA and glutamate in the CNS and NO production, is suggested to result in the diminished transmission of sympathetic tone to the periphery [101, 103].

Treatments with BPP-10c also induced an increase in AsS gene expression [22]. In contrast, nNOS was not found differentially expressed in the brains of SHRs treated with BPP-10c compared to vehicle-treated animals. On the other hand, the gene expression levels of eNOS, similarly to those of AsS, were found increased in the brains of SHRs animals treated with BPP-10c [22]. This data is in line with the results obtained by Kishi and colleagues [104] who were able to show that the overexpression of eNOS in the rostral ventrolateral medulla and the nucleus of the solitary tract of hypertensive rats results in reduced systolic arterial pressure and reduced HR.

However, the specificity of NO reactions with neuronal targets is determined in part by the precise localization of NOS within the cell. The targeting of NOS to discrete nuclei of neurons, mediated by adapter proteins, allowed to suggest that both synapsin and NOS participate of a ternary complex, which changes in the subcellular localization of NOS [105].

Knowing that the BPP-10c binds to synapsin in the CNS, we hypothesized the formation of a quaternary complex upon binding of BPP-10c with synapsin. The formation of this complex would direct the reactions of NO in neural targets, which would be determined in part by the location of this complex and by targeting NOS to specific sites of neurons [105].

We were able to show that BPP-10c is capable to induce intracellular Ca^{2+} signaling that involves the activation of GPCRs, NO production, and release of neurotransmitters, such as GABA and glutamate [22, 74]. The amino acid glutamate is the major excitatory neurotransmitter in the CNS of mammals, whereas GABA is the main mediator of sympathetic inhibitory currents. Both glutamate and GABA play key roles in the control of cardiovascular function in the CNS [103]. Excitatory amino acid neurotransmitters, like glutamate and aspartate, generally cause pressure responses and tachycardia, while inhibitory amino acid neurotrans-mitters, namely GABA and glycine, are responsible for depressing bradycardia [104]. It is well established that the excitatory amino acid glutamate is considered the main neurotransmitter of primary afferent fibers of baroreceptors to the nucleus tractus solitarii (NTS) [106]. Further-more, an excitatory projection from NTS to the caudal ventrolateral medulla (CVLM) is an essential part of the circuit of baroreflex control. The CVLM communicates with the rostral ventrolateral medulla (RVLM) by secretion of GABA. In addition to GABAergic inhibition of RVLM, excitatory amino acids are also known to exert important roles in cardiovascular regulation [107]. These neurotransmitters can regulate vasodilatation through reduction of both sympathetic activity and baroreflex sensitivity control, by acting on regulation of both sympathetic and parasympathetic systems. Therefore, the augmented baroreflex sensitivity by i.v. injection of BPP-10c is attributted to the release of these neurotransmitters [22, 74]. These data is summarized in Figure 8.

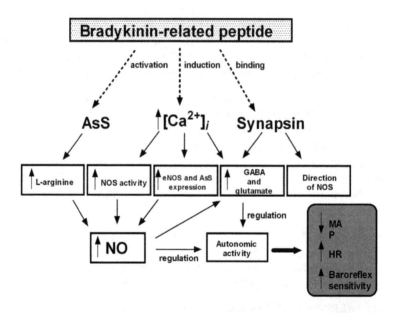

Figure 8. Schematic representation of BPP-10c mechanism of action in the CNS. According to [22, 74, 93], the pro-posed mechanism to explain the BPP-10c effects on the blood pressure (BP) and heart rate (HR) in spontaneously hy-pertensive rats (SHRs) was summarized in this figure. First, BPP-10c-induced [Ca^{+}] elevations activates signal transduction pathways responsible for the increased nitric oxide synthase (NOS) activity and the expression of the en-zymes, namely endothelial NOS (eNOS) and argininosuccinate synthase (ASS). It also triggers the release of the neuro-transmitters GABA and glutamate. After the BPP-10c internalization by neuronal cells, this peptide binds to synapsin to control the release of GABA and glutamate, and to direct the NOS to discrete cores of the neurons. Furthermore, BPP-10c can positively activate the AsS functions to increase the levels of L-arginine. NO production due to increased concentration of L-arginine, NOS activation, and increased expression of eNOS and AsS should contribute for the re-lease of neurotransmitters and also for the regulation of autonomic activity. Likewise GABA and glutamate deter-mine the reduction of both blood pressure and heart rate, and the increases of the baroreflex sensitivity in SHRs.

The fact of BPPs decrease the HR does not mean that its action is limited to the CNS. Taking the example of BPP-10c, *in vivo* biodistribution studies showed a significant presence of this peptide in the brain, however accumulation was also observed in the rat kidney [78]. Considering its high accumulation in kidneys with the fact that BPP-10c induces NO production in endothelial cells [20, 108] and the increase of plasma L-arginine level *in vivo* [20], we conclude that this peptide, and also potentially the other BPPs, may display both peripheral and central actions.

Moreover, there might exhist BPPs with exclusive peripheral action. As suggested for BPP-9a that promotes discret decrease of MAP and does not affect HR [75], whose effects were possible to be explained solely based on the classical mechanism of action suggested for the BPPs, *i.e.*, relying in the selective inhibition of the somatic ACE [73].

The pharmaceutical compositions for the applications in chornic-degenerative diseases and hypertension of the BPPs and their structural and/or conformational analogs, as well as the isolation and purification of BRPs secreted by snake venom glands are protected by the patents US20050031604 and US20080199503. The inventions further refer to pharmaceutical compositions that increase the biodisponibility and efficacy of BRPs peripheral and central biological activities. BRPs allowed the development a successful oral drug to treat human hypertension, but they also have the potential to become a drug by itself or a drug model to develop compouds devoted to treat central nervous system diseases, once pharmaceutical compositions that allow efficient delivery *in vivo* of these BRPs is achieved.

4.4. Mechanism of action that underlies the hyperalgesia and inflammatory responses

For many years it has been known that BK is an inflammatory mediator involved in the nociceptive process [109]. BK and also the BRPs produce pain and hyperalgesia due to their ability to excite and/or sensitize nociceptors [10].

In particular, two novel BRPs named fulvonin [SIVLRGKAPFR] and cyphokinin [DTRPPGFTPFR] were recently described in wasp (*Cyphononyx fulvognathus*). They could be structurally and functionally considered as BRPs, since they both are able to inhibit ACE as well as to induce the hyperalgesic effect in living rats after intraplantar injection, mostly due to the agonist action of these peptides on distinct B2 or B1 receptors, respectively [10].

5. Potential and effective pharmaceutical applications

5.1. Application of BPPs to treat CNS disorders and hypertension

In the last few years, as presented here, it was possible to describe a number of new mechanisms of action for BRPs previously known only as potent ACE inhibitors. Taking this into account, many pharmaceutical applications could be possible suggested for these peptides solely based on the treatment of pathologies related with their targets, for instace the somatic ACE, AsS and so on.

Regarding AsS as a novel potential target for the development of new drugs based on BPPs structural model, it is worth mentioning that its action on L-arginine metabolims contributes to three main functions in the organism, depending on the cell or tissue type involved, including effects on detoxification of ammonia in the liver, production of L-arginine in the kidney to be distributed to the whole organism, and the synthesis of L-arginine for the production of NO in several other cells [94]. In addition to these three major functions, it has also been suggested that AsS plays an important role in neuromodulation by producing argininosuccinate [110].

Due to its involvement in biochemical processes that generate physiological impacts on the organism, AsS is of great clinical value since its deficiency or excessive expression has been associated with some diseases, such as citrullinemia [111], hypertension [112, 113], and Alzheimer's disease [114, 115].

Since, AsS also participates in the L-arginine recycling, which contributes to the maintenance of NOS substrate,and AsS catalytic activity is considered the limiting step for NO production [116], the upregulation of this enzyme restores the balance of that system with consequent reduction in blood pressure [20].

Moreover, the identification of acetylcholine receptors as novel putative targets responsible for the vasodilatation promoted by the BRPs [19, 23, 75] also opens new avenues to the development of possible future therapeutic applications of BRPs related compounds for CNS diseases treatment.

Different experimental approaches demonstrate that acetylcholine muscarinic receptors are present in virtually all organs, tissues and cell types. The muscarinic receptors in the CNS are involved in the regulation of an extraordinary number of cognitive, behavioral, sensory, motor, and autonomous functions. Reduced or increased signalling of different subtypes of muscarinic receptors are involved in the pathophysiology of several diseases of the CNS including Alzheimer's and Parkinson's diseases, depression, schizophrenia, and epilepsy [117].

The contribution of the muscarinic acetylcholine receptor, mAchR-M1, in the NO production stimulated by BPP-5a is therapeutically and scientifically interesting, since much effort has been undertaken in the search for mAchR-M1 agonists to treat cognitive disorders including the Alzheimer's disease [118, 119].

On other hand, the mAchR-M3 is mainly involved in the control of vascular tone. The main actions mediated by these peripheral muscarinic receptors include the reduction of HR, stimulation of glandular secretion, and smooth muscle relaxation [117]. Compounds that activate this receptor to promote vasorelaxation, such as pilocarpine, are used for the treatment of glaucoma, ocular hypertension [120, 121]. However, the BRPs as BPP-13a is not only able to activate mAchR-M3, but in the same time it acts modulating the AsS activity. Since both pathways contribute to the antihypertensive effects by controlling the NO production, BPP-13a represents a potent vasodilator compound with potential broader applications in the medicine [75].

Taking together, the comparison of the biological actions of BRPs found in the venom and brain of the pit viper *Bothrops jararaca* with those from different species, allowed us to describe that, contrary to what was thought for several decades, these peptides have different biological effects and therefore are an inexhaustible source of powerful biological tools not only for the study and discovery of new physiological pathways, but also as potentially useful compounds for new drug development.

Therefore, a patent protecting the use of these oligopeptides capable of binding to diverse targets, determining the increase and the sustenance of nitric oxide (NO) production in mammalian cells by potentiating the endogenous argininosuccinate synthase activity present in animal cells and/or by increasing the intracellular bivalent free calcium ion in the cytosol of cells was filed (US20100035822). Pharmaceutical compositions containing one or more of these peptides is also described and also disclosed in this same patent.

5.2. Application of BIPs in the treatment of pain and inflammation

NO is involved in vasodilatation and in many other physiological processes. Several lines of evidence have indicated that NO plays a complex and diverse role in the modulation of pain [122]. It has been shown that NO mediates the analgesic effect of opioids and other analgesic substances, opening opportunities of potential use of molecules able to regulate NO production in pain therapy. Modification of pre-existing analgesic and anti-inflammatory drugs by addition of NO-releasing moieties has been shown to improve the analgesic efficacy of these drugs, and also to reduce their side effects [123].

NO donors have also been used with opioids to reduce pain in patients with cancer [124]. This strategy enhances the analgesic efficacy of morphine in patients with cancer pain, delaying the morphine tolerance and decreasing the incidence of the adverse effects of opioids [88, 125].

Nevertheless, the use of NO donors should be carefully evaluated, since the excessive levels of NO production can be deleterious to the organism [126]. Soon, keeping NO production at a safe level, avoiding the deleterious threshold, is of particular interest.

Therefore, the BRPs that modulate AsS activity [20, 75] could be considered for pain treatment. The fact that this AsS activity interferes with L-arginine metabolism, a source of NO synthesis that has its own levels subjected to very precise mechanisms of physiological control, represents the most likely target able to regulate NO production without generating undesired reactive by-products (review by [127]).

Other family of BRPs, which should be considered in the treatment of pain and inflammation, are the antagonists of BK receptors (BIPs), namely helokinestatins. Although these BRPs were described as vasoactive peptides, due to their ability to antagonize the relaxation effect induced by BK [35, 36], they could potencially be employed in the therapy of hyperalgesia.

Kinins formed following tissue trauma and in inflammatory processes, acting by means of the activation of B2 receptors, are among the most potent endogenous algogenic mediators. Kinins through action on BK receptors can release a large number of inflammatory mediators, such as prostaglandins and neuropeptides such as neurokinins [128], that in turn amplify the nociceptive response. Therefore, these receptors play an important role in pain transmission. BK produces a short-lived hyperalgesia, while des-Arg9-BK causes a long-lasting hyperalgesia [1].

However, most of the B2 receptor antagonists present partial agonist activity and fail to produce antinociception when given orally [1]. Non-peptide B2 receptor antagonists, although they are generally less potent when compared with Hoe 140, for instance, produce long-lasting oral antinociception with no evidence of partial agonistic activity [129-131]. In this way, further studies of BRPs applied in the pain treatment could provide valuable information for the development of novel peptidic or non-peptidic molecules to effectively relieve the pain of human patients.

The BRPs isolated from toad (*Bombina maxima*) defensive skin secretion, and their analogs thereof, prodrugs including the peptides, pharmaceutical compositions are protected by patent WO2004/068928. These BRPs and analogs thereof are antagonists of B2 receptor and they can be used to treat and/or prevent disorders associated with BK, including cardiovascular disorders, inflammation, asthma, allergic rhinitis, angiogenesis, pain and related pathologies.

Author details

Claudiana Lameu[1], Márcia Neiva[2] and Mirian A. F. Hayashi[2]

1 Departamento de Bioquímica, Instituto de Química, Universidade de São Paulo, Brazil

2 Departamento de Farmacologia, Escola Paulista de Medicina (EPM), Universidade Federal de São Paulo (UNIFESP), Brazil

References

[1] Calixto JB, Cabrini DA, Ferreira J, Campos MM. Kinins in pain and inflammation. Pain. 2000 Jul;87(1):1-5.

[2] Camargo AC, Ianzer D, Guerreiro JR, Serrano SM. Bradykinin-potentiating peptides: beyond captopril. Toxicon. 2012 Mar 15;59(4):516-23.

[3] Ferreira SH, Bartelt DC, Greene LJ. Isolation of bradykinin-potentiating peptides from Bothrops jararaca venom. Biochemistry. 1970 Jun 23;9(13):2583-93.

[4] Ferreira SH, Greene LH, Alabaster VA, Bakhle YS, Vane JR. Activity of various fractions of bradykinin potentiating factor against angiotensin I converting enzyme. Nature. 1970 Jan 24;225(5230):379-80.

[5] Rocha ESM, Beraldo WT, Rosenfeld G. Bradykinin, a hypotensive and smooth muscle stimulating factor released from plasma globulin by snake venoms and by trypsin. The American Journal of Physiology. 1949 Feb;156(2):261-73.

[6] Ondetti MA, Cushman DW. Design of protease inhibitors. Biopolymers. 1981 Sep; 20(9):2001-10.

[7] Hayashi MAF, Kerkis I. Toxinas como terapêuticos. São Paulo: Roca 2008.

[8] Bekheet SH, Awadalla EA, Salman MM, Hassan MK. Bradykinin potentiating factor isolated from Buthus occitanus venom has a protective effect against cadmium-induced rat liver and kidney damage. Tissue & Cell. 2011 Dec;43(6):337-43.

[9] Conceicao K, Konno K, de Melo RL, Antoniazzi MM, Jared C, Sciani JM, et al. Isolation and characterization of a novel bradykinin potentiating peptide (BPP) from the skin secretion of Phyllomedusa hypochondrialis. Peptides. 2007 Mar;28(3):515-23.

[10] Picolo G, Hisada M, Moura AB, Machado MF, Sciani JM, Conceicao IM, et al. Bradykinin-related peptides in the venom of the solitary wasp Cyphononyx fulvognathus. Biochemical Pharmacology. 2010 Feb 1;79(3):478-86.

[11] Zeng XC, Li WX, Peng F, Zhu ZH. Cloning and characterization of a novel cDNA sequence encoding the precursor of a novel venom peptide (BmKbpp) related to a bradykinin-potentiating peptide from Chinese scorpion Buthus martensii Karsch. IUBMB Life. 2000 Mar;49(3):207-10.

[12] Higuchi S, Murayama N, Saguchi K, Ohi H, Fujita Y, Camargo AC, et al. Bradykinin-potentiating peptides and C-type natriuretic peptides from snake venom. Immunopharmacology. 1999 Oct 15;44(1-2):129-35.

[13] Gomes CL, Konno K, Conceicao IM, Ianzer D, Yamanouye N, Prezoto BC, et al. Identification of novel bradykinin-potentiating peptides (BPPs) in the venom gland of a rattlesnake allowed the evaluation of the structure-function relationship of BPPs. Biochemical Pharmacology. 2007 Nov 1;74(9):1350-60.

[14] Hayashi MA, Murbach AF, Ianzer D, Portaro FC, Prezoto BC, Fernandes BL, et al. The C-type natriuretic peptide precursor of snake brain contains highly specific inhibitors of the angiotensin-converting enzyme. Journal of Neurochemistry. 2003 May;85(4):969-77.

[15] Murayama N, Hayashi MA, Ohi H, Ferreira LA, Hermann VV, Saito H, et al. Cloning and sequence analysis of a Bothrops jararaca cDNA encoding a precursor of seven bradykinin-potentiating peptides and a C-type natriuretic peptide. Proceedings of the National Academy of Sciences of the United States of America. 1997 Feb 18;94(4): 1189-93.

[16] Fernandez JJ, Li S. An improved algorithm for anisotropic nonlinear diffusion for denoising cryo-tomograms. Journal of Structural Biology. 2003 Oct-Nov;144(1-2): 152-61.

[17] Cho Y, Somer BG, Amatya A. Natriuretic peptides and their therapeutic potential. Heart Disease (Hagerstown, Md. 1999 Nov-Dec;1(5):305-28.

[18] Komatsu Y, Nakao K, Suga S, Ogawa Y, Mukoyama M, Arai H, et al. C-type natriuretic peptide (CNP) in rats and humans. Endocrinology. 1991 Aug;129(2):1104-6.

[19] Nery AA, Trujillo CA, Lameu C, Konno K, Oliveira V, Camargo AC, et al. A novel physiological property of snake bradykinin-potentiating peptides-reversion of MK-801 inhibition of nicotinic acetylcholine receptors. Peptides. 2008 Oct;29(10): 1708-15.

[20] Guerreiro JR, Lameu C, Oliveira EF, Klitzke CF, Melo RL, Linares E, et al. Argininosuccinate synthetase is a functional target for a snake venom anti-hypertensive peptide: role in arginine and nitric oxide production. The Journal of Biological Chemistry. 2009 Jul 24;284(30):20022-33.

[21] Ianzer D, Xavier CH, Fraga FC, Lautner RQ, Guerreiro JR, Machado LT, et al. BPP-5a produces a potent and long-lasting NO-dependent antihypertensive effect. Therapeutic Advances in Cardiovascular Disease. 2011 Dec;5(6):281-95.

[22] Lameu C, Pontieri V, Guerreiro JR, Oliveira EF, da Silva CA, Giglio JM, et al. Brain nitric oxide production by a proline-rich decapeptide from Bothrops jararaca venom improves baroreflex sensitivity of spontaneously hypertensive rats. Hypertens Res. 2010 Dec;33(12):1283-8.

[23] Morais KL, Hayashi MA, Bruni FM, Lopes-Ferreira M, Camargo AC, Ulrich H, et al. Bj-PRO-5a, a natural angiotensin-converting enzyme inhibitor, promotes vasodilatation mediated by both bradykinin B(2)and M1 muscarinic acetylcholine receptors. Biochemical Pharmacology. 2011 Mar 15;81(6):736-42.

[24] Tanen DA, Ruha AM, Graeme KA, Curry SC, Fischione MA. Rattlesnake envenomations: unusual case presentations. Archives of Internal Medicine. 2001 Feb 12;161(3): 474-9.

[25] Walter FG, Bilden EF, Gibly RL. Envenomations. Critical care clinics. 1999 Apr;15(2): 353-86, ix.

[26] Ferreira SH. A Bradykinin-Potentiating Factor (BPF) Present in the venom of Bothrops jararaca. British Journal of Pharmacology and Chemotherapy. 1965 Feb; 24:163-9.

[27] Camargo AC, Graeff FG. Subcellular distribution and properties of the bradykinin inactivation system in rabbit brain homogenates. Biochemical Pharmacology. 1969 Feb;18(2):548-9.

[28] Ribeiro SA, Corrado AP, Graeff FG. Antinociceptive action of intraventricular bradykinin. Neuropharmacology. 1971 Nov;10(6):725-31.

[29] Hayashi MA, Camargo AC. The Bradykinin-potentiating peptides from venom gland and brain of Bothrops jararaca contain highly site specific inhibitors of the somatic angiotensin-converting enzyme. Toxicon. 2005 Jun 15;45(8):1163-70.

[30] Soares MR, Oliveira-Carvalho AL, Wermelinger LS, Zingali RB, Ho PL, Junqueira-de-Azevedo IL, et al. Identification of novel bradykinin-potentiating peptides and C-type natriuretic peptide from Lachesis muta venom. Toxicon. 2005 Jul;46(1):31-8.

[31] Murayama N, Michel GH, Yanoshita R, Samejima Y, Saguchi K, Ohi H, et al. cDNA cloning of bradykinin-potentiating peptides - C-type natriuretic peptide precursor, and characterization of the novel peptide Leu3-blomhotin from the venom of Agkistrodon blomhoffi. European Journal of Biochemistry / FEBS. 2000 Jul;267(13):4075-80.

[32] Ianzer D, Konno K, Marques-Porto R, Vieira Portaro FC, Stocklin R, Martins de Camargo AC, et al. Identification of five new bradykinin potentiating peptides (BPPs) from Bothrops jararaca crude venom by using electrospray ionization tandem mass spectrometry after a two-step liquid chromatography. Peptides. 2004 Jul;25(7): 1085-92.

[33] Cheung HS, Cushman DW. Inhibition of homogeneous angiotensin-converting enzyme of rabbit lung by synthetic venom peptides of Bothrops jararaca. Biochimica et Biophysica Acta. 1973 Feb 15;293(2):451-63.

[34] Ma C, Yang M, Zhou M, Wu Y, Wang L, Chen T, et al. The natriuretic peptide/helokinestatin precursor from Mexican beaded lizard (Heloderma horridum) venom: Amino acid sequence deduced from cloned cDNA and identification of two novel encoded helokinestatins. Peptides. 2011 Jun;32(6):1166-71.

[35] Kwok HF, Chen T, O'Rourke M, Ivanyi C, Hirst D, Shaw C. Helokinestatin: a new bradykinin B2 receptor antagonist decapeptide from lizard venom. Peptides. 2008 Jan;29(1):65-72.

[36] Zhang Y, Wang L, Zhou M, Zhou Z, Chen X, Chen T, et al. The structure of helokinestatin-5 and its biosynthetic precursor from Gila monster (Heloderma suspectum) venom: evidence for helokinestatin antagonism of bradykinin-induced relaxation of rat tail artery smooth muscle. Peptides. 2010 Aug;31(8):1555-61.

[37] Fry BG, Roelants K, Winter K, Hodgson WC, Griesman L, Kwok HF, et al. Novel venom proteins produced by differential domain-expression strategies in beaded lizards and gila monsters (genus Heloderma). Molecular Biology and Evolution. 2010 Feb;27(2):395-407.

[38] Fry BG, Winter K, Norman JA, Roelants K, Nabuurs RJ, van Osch MJ, et al. Functional and structural diversification of the Anguimorpha lizard venom system. Molecular Cell Proteomics. 2010 Nov;9(11):2369-90.

[39] Higuchi S, Murayama N, Saguchi K, Ohi H, Fujita Y, da Silva NJ, Jr., et al. A novel peptide from the ACEI/BPP-CNP precursor in the venom of Crotalus durissus collili-

neatus. Comparative Biochemistry and Physiology - Part C: Toxicology & Pharmacology. 2006 Oct;144(2):107-21.

[40] Fry BG, Vidal N, Norman JA, Vonk FJ, Scheib H, Ramjan SF, et al. Early evolution of the venom system in lizards and snakes. Nature. 2006 Feb 2;439(7076):584-8.

[41] Irwin DM. cDNA cloning of proglucagon from the stomach and pancreas of the dog. DNA Sequence. 2001 Nov;12(4):253-60.

[42] Ohno M, Menez R, Ogawa T, Danse JM, Shimohigashi Y, Fromen C, et al. Molecular evolution of snake toxins: is the functional diversity of snake toxins associated with a mechanism of accelerated evolution? Progress in Nucleic Acid Research and Molecular Biology. 1998;59:307-64.

[43] Takei Y. Structural and functional evolution of the natriuretic peptide system in vertebrates. International Review of Cytology. 2000;194:1-66.

[44] Sanz L, Ayvazyan N, Calvete JJ. Snake venomics of the Armenian mountain vipers Macrovipera lebetina obtusa and Vipera raddei. Journal of Proteomics. 2008 Jul 21;71(2):198-209.

[45] Schweitz H, Vigne P, Moinier D, Frelin C, Lazdunski M. A new member of the natriuretic peptide family is present in the venom of the green mamba (Dendroaspis angusticeps). The Journal of Biological Chemistry. 1992 Jul 15;267(20):13928-32.

[46] Nakashima K, Nobuhisa I, Deshimaru M, Nakai M, Ogawa T, Shimohigashi Y, et al. Accelerated evolution in the protein-coding regions is universal in crotalinae snake venom gland phospholipase A2 isozyme genes. Proceedings of the National Academy of Sciences of the United States of America. 1995 Jun 6;92(12):5605-9.

[47] Ho PL, Soares MB, Maack T, Gimenez I, Puorto G, Furtado MF, et al. Cloning of an unusual natriuretic peptide from the South American coral snake Micrurus corallinus. European Journal of Biochemistry / FEBS. 1997 Nov 15;250(1):144-9.

[48] Junqueira de Azevedo IL, Farsky SH, Oliveira ML, Ho PL. Molecular cloning and expression of a functional snake venom vascular endothelium growth factor (VEGF) from the Bothrops insularis pit viper. A new member of the VEGF family of proteins. The Journal of Biological Chemistry. 2001 Oct 26;276(43):39836-42.

[49] Eddy SR. Non-coding RNA genes and the modern RNA world. Nature Reviews. 2001 Dec;2(12):919-29.

[50] Mighell AJ, Smith NR, Robinson PA, Markham AF. Vertebrate pseudogenes. FEBS Letters. 2000 Feb 25;468(2-3):109-14.

[51] Hirotsune S. [An expressed pseudogene regulates mRNA stability of its homologous coding gene]. Tanpakushitsu Kakusan Koso. 2003 Nov;48(14):1908-12.

[52] Mignone F, Pesole G. rRNA-like sequences in human mRNAs. Applied Bioinformatics. 2002;1(3):145-54.

[53] Greene LJ, Camargo AC, Krieger EM, Stewart JM, Ferreira SH. Inhibition of the con-
      version of angiotensin I to II and potentiation of bradykinin by small peptides
      present in Bothrops jararaca venom. Circulation Research. 1972 Sep;31(9):Suppl
      2:62-71.

[54] Gavras H, Brunner HR, Laragh JH, Sealey JE, Gavras I, Vukovich RA. An angiotensin
      converting-enzyme inhibitor to identify and treat vasoconstrictor and volume factors
      in hypertensive patients. The New England Journal of Medicine. 1974 Oct 17;291(16):
      817-21.

[55] Gavras H, Gavras I, Textor S, Volicer L, Brunner HR, Rucinska EJ. Effect of angioten-
      sin converting enzyme inhibition on blood pressure, plasma renin activity and plas-
      ma aldosterone in essential hypertension. The Journal of Clinical Endocrinology and
      Metabolism. 1978 Feb;46(2):220-6.

[56] Gavras H, Brunner HR, Laragh JH, Gavras I, Vukovich RA. The use of angiotensin-
      converting enzyme inhibitor in the diagnosis and treatment of hypertension. Clinical
      Science and Molecular Medicine. 1975 Jun;2:57s-60s.

[57] Krieger EM, Salgado HC, Assan CJ. Greene LL, Ferreira SH: Potential screening test
      for detection of overactivity of renin-angiotensin system. Lancet. 1971 Feb 6;1(7693):
      269-71.

[58] Stewart JM, Ferreira SH, Greene LJ. Bradykinin potentiating peptide PCA-Lys-Trp-
      Ala-Pro. An inhibitor of the pulmonary inactivation of bradykinin and conversion of
      angiotensin I to II. Biochemical Pharmacology. 1971 Jul;20(7):1557-67.

[59] Cushman DW, Cheung HS, Sabo EF, Ondetti MA. Design of potent competitive in-
      hibitors of angiotensin-converting enzyme. Carboxyalkanoyl and mercaptoalkanoyl
      amino acids. Biochemistry. 1977 Dec 13;16(25):5484-91.

[60] Turner AJ, Hooper NM. The angiotensin-converting enzyme gene family: genomics
      and pharmacology. Trends in Pharmacological Sciences. 2002 Apr;23(4):177-83.

[61] Acharya KR, Sturrock ED, Riordan JF, Ehlers MR. Ace revisited: a new target for
      structure-based drug design. Nat Rev Drug Discov. 2003 Nov;2(11):891-902.

[62] Kurtz A, Wagner C. Cellular control of renin secretion. The Journal of Experimental
      Biology. 1999 Feb;202(Pt 3):219-25.

[63] Ben-Ari ET, Garrison JC. Regulation of angiotensinogen mRNA accumulation in rat
      hepatocytes. The American Journal of Physiology. 1988 Jul;255(1 Pt 1):E70-9.

[64] Ng KK, Vane JR. Conversion of angiotensin I to angiotensin II. Nature. 1967 Nov
      25;216(5117):762-6.

[65] de Gasparo M, Catt KJ, Inagami T, Wright JW, Unger T. International union of phar-
      macology. XXIII. The angiotensin II receptors. Pharmacological Reviews. 2000 Sep;
      52(3):415-72.

[66] Moreau ME, Garbacki N, Molinaro G, Brown NJ, Marceau F, Adam A. The kallikrein-kinin system: current and future pharmacological targets. Journal of Pharmacological Sciences. 2005 Sep;99(1):6-38.

[67] Margolius HS. Theodore Cooper Memorial Lecture. Kallikreins and kinins. Some unanswered questions about system characteristics and roles in human disease. Hypertension. 1995 Aug;26(2):221-9.

[68] Marceau F, Bachvarov DR. Kinin receptors. Clinical Reviews in Allergy & Immunology. 1998 Winter;16(4):385-401.

[69] Wei L, Alhenc-Gelas F, Corvol P, Clauser E. The two homologous domains of human angiotensin I-converting enzyme are both catalytically active. The Journal of Biological Chemistry. 1991 May 15;266(14):9002-8.

[70] Jaspard E, Wei L, Alhenc-Gelas F. Differences in the properties and enzymatic specificities of the two active sites of angiotensin I-converting enzyme (kininase II). Studies with bradykinin and other natural peptides. The Journal of Biological Chemistry. 1993 May 5;268(13):9496-503.

[71] Rousseau A, Michaud A, Chauvet MT, Lenfant M, Corvol P. The hemoregulatory peptide N-acetyl-Ser-Asp-Lys-Pro is a natural and specific substrate of the N-terminal active site of human angiotensin-converting enzyme. The Journal of Biological Chemistry. 1995 Feb 24;270(8):3656-61.

[72] Camargo A, Ferreira SH. Action of bradykinin potentiating factor (BPF) and dimercaprol (BAL) on the responses to bradykinin of isolated preparations of rat intestines. British Journal of Pharmacology. 1971 Jun;42(2):305-7.

[73] Cotton J, Hayashi MA, Cuniasse P, Vazeux G, Ianzer D, De Camargo AC, et al. Selective inhibition of the C-domain of angiotensin I converting enzyme by bradykinin potentiating peptides. Biochemistry. 2002 May 14;41(19):6065-71.

[74] Lameu C, Hayashi MA, Guerreiro JR, Oliveira EF, Lebrun I, Pontieri V, et al. The central nervous system as target for antihypertensive actions of a proline-rich peptide from Bothrops jararaca venom. Cytometry A. 2010 Mar;77(3):220-30.

[75] Morais KLP, Ianzer D, Santos RAS, Miranda JRR, Melo RL, Guerreiro JR, et al. The structural diversity of proline-rich oligopeptides from Bothrops jararaca (Bj-PROs) provides synergistic cardiovascular actions Hypertension Research. submitted.

[76] Mueller S, Gothe R, Siems WD, Vietinghoff G, Paegelow I, Reissmann S. Potentiation of bradykinin actions by analogues of the bradykinin potentiating nonapeptide BPP9alpha. Peptides. 2005 Jul;26(7):1235-47.

[77] Ianzer D, Santos RA, Etelvino GM, Xavier CH, de Almeida Santos J, Mendes EP, et al. Do the cardiovascular effects of angiotensin-converting enzyme (ACE) I involve ACE-independent mechanisms? new insights from proline-rich peptides of Bothrops

jararaca. The Journal of Pharmacology and Experimental Therapeutics. 2007 Aug; 322(2):795-805.

[78] Silva CA, Portaro FC, Fernandes BL, Ianzer DA, Guerreiro JR, Gomes CL, et al. Tissue distribution in mice of BPP 10c, a potent proline-rich anti-hypertensive peptide of Bothrops jararaca. Toxicon. 2008 Mar 15;51(4):515-23.

[79] Flam BR, Eichler DC, Solomonson LP. Endothelial nitric oxide production is tightly coupled to the citrulline-NO cycle. Nitric Oxide. 2007 Nov-Dec;17(3-4):115-21.

[80] Shen LJ, Beloussow K, Shen WC. Accessibility of endothelial and inducible nitric oxide synthase to the intracellular citrulline-arginine regeneration pathway. Biochemical Pharmacology. 2005 Jan 1;69(1):97-104.

[81] Ignarro LJ, Byrns RE, Wood KS. Endothelium-dependent modulation of cGMP levels and intrinsic smooth muscle tone in isolated bovine intrapulmonary artery and vein. Circulation Research. 1987 Jan;60(1):82-92.

[82] Finkbeiner S, Greenberg ME. Spatial features of calcium-regulated gene expression. Bioessays. 1997 Aug;19(8):657-60.

[83] Bianchi A, Evans DB, Cobb M, Peschka MT, Schaeffer TR, Laffan RJ. Inhibition by SQ 20881 of vasopressor response to angiotensin I in conscious animals. European Journal of Pharmacology. 1973 Jul;23(1):90-6.

[84] Bunag RD, Walaszek EJ, Mueting N. Sex differences in reflex tachycardia induced by hypotensive drugs in unanesthetized rats. The American Journal of Physiology. 1975 Sep;229(3):652-6.

[85] Krieger EM. Neurogenic Hypertension in the Rat. Circulation research. 1964 Dec; 15:511-21.

[86] Chapleau MW, Li Z, Meyrelles SS, Ma X, Abboud FM. Mechanisms determining sensitivity of baroreceptor afferents in health and disease. Annals of the New York Academy of Sciences. 2001 Jun;940:1-19.

[87] Michelini LC. Regulação neuro-humoral da pressão arterial. Rio de Janeiro: Ghuanabar Kogan 1999.

[88] Lanfranchi PA, Somers VK. Arterial baroreflex function and cardiovascular variability: interactions and implications. American Journal of Physiology. 2002 Oct; 283(4):R815-26.

[89] Irigoyen MC, Moreira ED, Werner A, Ida F, Pires MD, Cestari IA, et al. Aging and baroreflex control of RSNA and heart rate in rats. American Journal of Physiology. 2000 Nov;279(5):R1865-71.

[90] Schlaich MP, Lambert E, Kaye DM, Krozowski Z, Campbell DJ, Lambert G, et al. Sympathetic augmentation in hypertension: role of nerve firing, norepinephrine reuptake, and Angiotensin neuromodulation. Hypertension. 2004 Feb;43(2):169-75.

[91] Velden M, Karemaker JM, Wolk C, Schneider R. Inferring vagal effects on the heart from changes in cardiac cycle length: implications for cycle time-dependency. Int J Psychophysiol. 1990 Nov;10(1):85-93.

[92] Somsen RJ, Jennings JR, Van der Molen MW. The cardiac cycle time effect revisited: temporal dynamics of the central-vagal modulation of heart rate in human reaction time tasks. Psychophysiology. 2004 Nov;41(6):941-53.

[93] Lameu C. The central nervous system as target for anti-hypertensive actions of a pro-line-rich peptide from Bothrops jararaca venom [PhD thesis, Tese para obtenção do título doutor em ciências (bioquímica)]. São Paulo: Universidade de São Paulo; 2009.

[94] Husson A, Brasse-Lagnel C, Fairand A, Renouf S, Lavoinne A. Argininosuccinate synthetase from the urea cycle to the citrulline-NO cycle. European Journal of Biochemistry / FEBS. 2003 May;270(9):1887 99.

[95] Bredt DS, Snyder SH. Isolation of nitric oxide synthetase, a calmodulin-requiring enzyme. Proceedings of the National Academy of Sciences of the United States of America. 1990 Jan;87(2):682-5.

[96] Sears CE, Ashley EA, Casadei B. Nitric oxide control of cardiac function: is neuronal nitric oxide synthase a key component? Philosophical transactions of the Royal Society of London. 2004 Jun 29;359(1446):1021-44.

[97] Umans JG, Levi R. Nitric oxide in the regulation of blood flow and arterial pressure. Annual Review of Physiology. 1995;57:771-90.

[98] Bolotina VM, Najibi S, Palacino JJ, Pagano PJ, Cohen RA. Nitric oxide directly activates calcium-dependent potassium channels in vascular smooth muscle. Nature. 1994 Apr 28;368(6474):850-3.

[99] Murphy ME, Brayden JE. Nitric oxide hyperpolarizes rabbit mesenteric arteries via ATP-sensitive potassium channels. The Journal of Physiology. 1995 Jul 1;486 (Pt 1): 47-58.

[100] Garthwaite J, Boulton CL. Nitric oxide signaling in the central nervous system. Annual Review of Physiology. 1995;57:683-706.

[101] Patel KP, Li YF, Hirooka Y. Role of nitric oxide in central sympathetic outflow. Experimental Biology and Medicine (Maywood, NJ. 2001 Oct;226(9):814-24.

[102] Pontieri V, Venezuela MK, Scavone C, Michelini LC. Role of endogenous nitric oxide in the nucleus tratus solitarii on baroreflex control of heart rate in spontaneously hypertensive rats. Journal of Hypertension. 1998 Dec;16(12 Pt 2):1993-9.

[103] Gordon FJ, Sved AF. Neurotransmitters in central cardiovascular regulation: glutamate and GABA. Clinical and Experimental Pharmacology & Physiology. 2002 May-Jun;29(5-6):522-4.

[104] Kishi T, Hirooka Y, Sakai K, Shigematsu H, Shimokawa H, Takeshita A. Overexpression of eNOS in the RVLM causes hypotension and bradycardia via GABA release. Hypertension. 2001 Oct;38(4):896-901.

[105] Jaffrey SR, Erdjument-Bromage H, Ferris CD, Tempst P, Snyder SH. Protein S-nitrosylation: a physiological signal for neuronal nitric oxide. Nature Cell Biology. 2001 Feb;3(2):193-7.

[106] Lawrence AJ, Jarrott B. Neurochemical modulation of cardiovascular control in the nucleus tractus solitarius. Progress in Neurobiology. 1996 Jan;48(1):21-53.

[107] Ito S, Sved AF. Tonic glutamate-mediated control of rostral ventrolateral medulla and sympathetic vasomotor tone. The American Journal of Physiology. 1997 Aug; 273(2 Pt 2):R487-94.

[108] Benedetti G, Morais KL, Guerreiro JR, de Oliveira EF, Hoshida MS, Oliveira L, et al. Bothrops jararaca peptide with anti-hypertensive action normalizes endothelium dysfunction involved in physiopathology of preeclampsia. PloS One. 2011;6(8):e23680.

[109] Dray A, Perkins M. Bradykinin and inflammatory pain. Trends in Neurosciences. 1993 Mar;16(3):99-104.

[110] Nakamura H, Saheki T, Ichiki H, Nakata K, Nakagawa S. Immunocytochemical localization of argininosuccinate synthetase in the rat brain. The Journal of Comparative Neurology. 1991 Oct 22;312(4):652-79.

[111] Curis E, Nicolis I, Moinard C, Osowska S, Zerrouk N, Benazeth S, et al. Almost all about citrulline in mammals. Amino Acids. 2005 Nov;29(3):177-205.

[112] Dusse LM, Lwaleed BA, Silva RM, Cooper AJ, Carvalho MG. Nitric oxide in preeclampsia: be careful with the results! European Journal of Obstetrics, Gynecology, and Reproductive Biology. 2008 Jun;138(2):242-3.

[113] Panza JA. Endothelial dysfunction in essential hypertension. Clinical Cardiology. 1997 Nov;20(11 Suppl 2):II-26-33.

[114] Haas J, Storch-Hagenlocher B, Biessmann A, Wildemann B. Inducible nitric oxide synthase and argininosuccinate synthetase: co-induction in brain tissue of patients with Alzheimer's dementia and following stimulation with beta-amyloid 1-42 in vitro. Neuroscience Letters. 2002 Apr 5;322(2):121-5.

[115] Wiesinger H. Arginine metabolism and the synthesis of nitric oxide in the nervous system. Progress in Neurobiology. 2001 Jul;64(4):365-91.

[116] Solomonson LP, Flam BR, Pendleton LC, Goodwin BL, Eichler DC. The caveolar nitric oxide synthase/arginine regeneration system for NO production in endothelial cells. The Journal of Experimental Biology. 2003 Jun;206(Pt 12):2083-7.

[117]   Wess J. Muscarinic acetylcholine receptor knockout mice: novel phenotypes and clinical implications. Annual Review of Pharmacology and Toxicology. 2004;44:423-50.

[118]   Conn PJ, Jones CK, Lindsley CW. Subtype-selective allosteric modulators of muscarinic receptors for the treatment of CNS disorders. Trends in Pharmacological Sciences. 2009 Mar;30(3):148-55.

[119]   Fisher A, Pittel Z, Haring R, Bar-Ner N, Kliger-Spatz M, Natan N, et al. M1 muscarinic agonists can modulate some of the hallmarks in Alzheimer's disease: implications in future therapy. J Mol Neurosci. 2003;20(3):349-56.

[120]   Costagliola C, dell'Omo R, Romano MR, Rinaldi M, Zeppa L, Parmeggiani F. Pharmacotherapy of intraocular pressure: part I. Parasympathomimetic, sympathomimetic and sympatholytics. Expert opinion on Pharmacotherapy. 2009 Nov;10(16): 2663-77.

[121]   Hurvitz LM, Kaufman PL, Robin AL, Weinreb RN, Crawford K, Shaw B. New developments in the drug treatment of glaucoma. Drugs. 1991 Apr;41(4):514-32.

[122]   Cury Y, Picolo G, Gutierrez VP, Ferreira SH. Pain and analgesia: The dual effect of nitric oxide in the nociceptive system. Nitric Oxide. 2011 Oct 30;25(3):243-54.

[123]   Stefano F, Distrutti E. Cyclo-oxygenase (COX) inhibiting nitric oxide donating (CINODs) drugs: a review of their current status. Current topics in Medicinal Chemistry. 2007;7(3):277-82.

[124]   Toda N, Kishioka S, Hatano Y, Toda H. Modulation of opioid actions by nitric oxide signaling. Anesthesiology. 2009 Jan;110(1):166-81.

[125]   Lauretti GR, Lima IC, Reis MP, Prado WA, Pereira NL. Oral ketamine and transdermal nitroglycerin as analgesic adjuvants to oral morphine therapy for cancer pain management. Anesthesiology. 1999 Jun;90(6):1528-33.

[126]   Ridnour LA, Thomas DD, Mancardi D, Espey MG, Miranda KM, Paolocci N, et al. The chemistry of nitrosative stress induced by nitric oxide and reactive nitrogen oxide species. Putting perspective on stressful biological situations. Biological Chemistry. 2004 Jan;385(1):1-10.

[127]   Lameu C, de Camargo AC, Faria M. L-arginine signalling potential in the brain: the peripheral gets central. Recent Patents on CNS Drug Discovery. 2009 Jun;4(2):137-42.

[128]   Dray A, Perkins M. Kinins and pain.: London: Academic Press 1997.

[129]   Burgess GM, Perkins MN, Rang HP, Campbell EA, Brown MC, McIntyre P, et al. Bradyzide, a potent non-peptide B(2) bradykinin receptor antagonist with long-lasting oral activity in animal models of inflammatory hyperalgesia. British Journal of Pharmacology. 2000 Jan;129(1):77-86.

[130]   de Campos RO, Alves RV, Ferreira J, Kyle DJ, Chakravarty S, Mavunkel BJ, et al. Oral antinociception and oedema inhibition produced by NPC 18884, a non-peptidic

bradykinin B2 receptor antagonist. Naunyn-Schmiedeberg's Archives of Pharmacology. 1999 Sep;360(3):278-86.

[131] Griesbacher T, Amann R, Sametz W, Diethart S, Juan H. The nonpeptide B2 receptor antagonist FR173657: inhibition of effects of bradykinin related to its role in nociception. British Journal of Pharmacology. 1998 Jul;124(6):1328-34.

[132] Burnett JC, Jr. Vasopeptidase inhibition: a new concept in blood pressure management. J Hypertens Suppl. 1999 Feb;17(1):S37-43.

[133] Couture R, Girolami JP. Putative roles of kinin receptors in the therapeutic effects of angiotensin 1-converting enzyme inhibitors in diabetes mellitus. European Journal of Pharmacology. 2004 Oct 1;500(1-3):467-85.

[134] Santos RA, Frezard F, Ferreira AJ. Angiotensin-(1-7): blood, heart, and blood vessels. Current Medicinal Chemistry. 2005 Oct;3(4):383-91.

Chapter 7

Serine proteases — Cloning, Expression and Potential Applications

Camila Miyagui Yonamine, Álvaro Rossan de Brandão Prieto da Silva and Geraldo Santana Magalhães

Additional information is available at the end of the chapter

1. Introduction

1.1. Snake venom serine proteases

Serine proteases have been isolated from the venoms of viperidae snakes [1, 2] and affect several physiological processes such as the coagulation cascade. These enzymes are called snake venom serine proteases (SVSPs), they are multi-functional proteins with a catalytic triad formed by HDS amino acids [3].

The SVSPs resembles at least in part thrombin, a multifunctional protease that plays a key role in coagulation. Therefore these enzymes are denominated snake venom thrombin-like enzymes (SVTLEs), and are widely distributed in the venoms of several genera [4,5]. While thrombin is able to cleave both fibrinopeptide A (FPA) and fibrinopeptide B (FPB) from fibrinogen leading the formation of fibrin and activating factor XIII, some actions of SVTLEs usually cleave FPA alone and only a few cleave FPB. Thus, without cleavage of both FPA and FPB they are unable to activate factor XIII producing fibrin monomers that are not cross-linked, leading to clots markedly susceptible to digestion by plasmin and are rapidly removed from circulation by either reticuloendothelial phagocitosis and/or normal fibrinolysis. This process causes a breakdown in the fibrinolytic system and effective removal of fibrinogen from the plasma [6].

2. Body

There are three groups of snake venom fibrinogen clotting enzymes based on the rates of release of fibrinopeptides A and/or B from fibrinogen. In addition to SVLTEs, other SVSPs groups are active in other parts of the coagulation cascade, such as kallikrein-like enzymes (KN); plasminogen activators (PA); protein C like enzyme and factor V activators. One group releases fibrinopeptide A preferentially (the venombin A group including ancrod from venom of the Malayan pit viper, *Calloselasma rhodostoma*); another group releases both fibrinopeptides A and B (the venombin AB group including halystase and calobin from *Agkistrodon halys blomhoffii* and *Agkistrodon caliginosus*, respectively) and the third group releases fibrinopeptide B preferentially (the venombin B group including v enzyme from venom of the southern copperhead, *Agkistrodon contortrix contortrix*) [5,7,8]. Figure 1 summarizes some snake toxins that affect the blood coagulation cascade, based on [6].

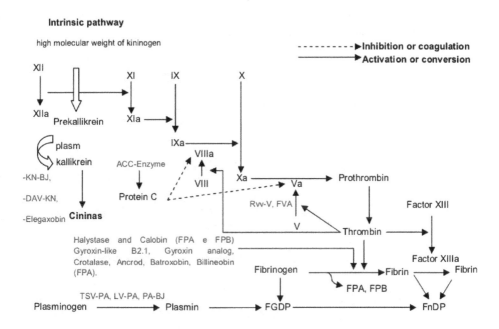

Figure 1. Some SVSPs acting in blood system. FGDP: Fibrinogen degradation products; FNDP: Fibrin degradation products; FPA: Fibrinopeptide A; FPB: Fibrinopeptide B. **KN-BJ,** *Bothrops jararaca* (O13069) [9]; **Dav-KN,** *Agkistrodon acutus* (Q9I8X0) [10]; **Elegaxobin-1,** *Trimeresurus elegans* (P84788) [11]; **ACC-C Protein C activator;** *Agkistrodon contortrix contortrix* (P09872) [12]; **RVV-Va** Russel's viper venom FV activator alpha, *Daboia russelli siamensis* (P18964) [13]; **FVA Factor V-activating enzyme,** *Vipera lebetina* (Q9PT41) [14]; **Halystase,** *Agkistrodon halys blomhoffii* (P81176) [15]; **Calobin,** *Agkistrodon caliginosus* (Q91053) [16]; **Gyroxin-like B2.1,** *Crotalus durissus terrificus* (Q58G94) [17]; **Gyroxin analog,** *Lachesis muta muta* (P33589) [18]; **Crotalase,** *Crotalus adamanteus* [19]; **Ancrod,** *Agkistrodon rhodostoma* (P26324) [20]; **Batroxobin,** *Bothrops atrox* (P04971) [21]; **Bilineobin,** *Agkistrodon bilineatus* (Q9PSN3) [22];**TSV-PA,** *Trimeresurus stejnegeri* (Q91516) [23]; **LV-PA,** *Lachesis muta muta* (P84036) [24]; **PA-BJ,** *Bothrops jararaca* (P81824) [25]. Toxin names were indicated in bold followed by snake species in italic and the Swissprot accession numbers were represented in parenthesis.

The major symptoms from snakebite affecting the haemostatic system are: (a) reduced coagulability of blood, resulting in an increased tendency to bleed, (b) bleeding due to the damage to blood vessels, (c) secondary effects of increased bleeding, ranging from hypovolaemic shock to secondary-organ damage, such as intracerebral haemorrhage, anterior pituitary haemorrhage or renal damage and (d) direct pathologic thrombosis and its sequelae, particularly pulmonary embolism [26].

The venom fibrinogenolytic serine proteases as well as the venom plasminogen activator, share extensive sequence homology with the thrombin-like venom serine proteases [27] such as ancrod [20, 28], batroxobin [21, 29], crotalase [30,31], gyroxin-like serine proteases [17], kallikrein-like enzyme from *Crotalus atrox* [32] and the protein C activator from *A. c. contortrix* [33] venoms.

The SVSPs share the conserved catalytic triad formed by the amino acids His, Asp and Ser, six disulfide bonds and global highly similarities, as seen in Figure 2. In this alignment, the deduced amino amino acid sequences of gyroxin-like B2.1, B1.3, B1.4, and B1.7 are highly similar to other SVSPs with several biological functions [17].

In order to predict the biological function of SVSPs, a functional dendrogram was generated based on the amino acid sequence alignment from Figure 2. Clearly there are subtle differences among these homologous enzymes that may explain different functions such as: fibrinogenolytic (group I), Aα fibrinogenases (subgroup I a), Protein C activator and CPI-enzyme (subgroup I b), kininogenases (subgroup I c), plasminogen activator (group II), factor V activators (group III) thrombin-like, or other specific enzymatic activities (Figure 3) [17].

Despite significant sequence identity (50–70%), SVSPs display high specificity toward distinct macromolecular substrates. Based on their biological roles, they have been classified as activators of the fibrinolytic system, procoagulant, anticoagulant and platelet-aggregating enzymes [34]. The procoagulant SVSPs activate FVII, FX and prothrombin [35] and shorten the coagulation times, while the anticoagulants inactivates factors Va and VIIIa and plays a key role in controlling haemostasis, Ancrod (from the Malayan pit viper, *Calloselasma rhodostoma*), in particular, has been used as an anticoagulant to achieve "therapeutic defibrination" [34].

As it can be seen in Figure 4 (top), the 3D model of gyroxin-like B2.1 shows the catalytic site (Ser184, His43 and Asp88) superimposed with the catalytic site of thrombin (Ser$_{195}$, His$_{57}$ and Asp$_{102}$) [36]. The overall structure (bottom) show the typical fold of a serine proteinase in which the active-site cleft is located at the junction of the two six-stranded β-barrels. Among the conserved 3D structural features between trypsin-like enzyme and SVSPs are the two β-barrel subdomains, the orientation of catalytic site and the pattern of Cys residues. In contrast with other serine proteases, a unique long C-terminal tail of gyroxin are highly conserved only on SVSPs. In addition, SVSPs are active only as a single chain enzyme while prothrombin is activated by Factor Xa generating the Light (L) and Heavy (H) chains of active thrombin.

2.1. The role of Protease Activated Receptor (PAR) on serine protease coagulation

Protease-activated receptors (PARs) are members of family of seven-transmembrane G-protein-coupled receptors (GPCRs). The activation is triggered by the cleavage of the N-terminus of the receptor by a serine protease, resulting in the generation of a new tethered ligand that interacts with the receptor within its extracellular loop-2. This ligand binding to

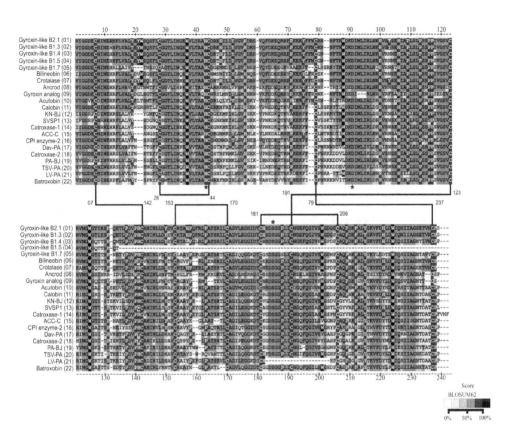

Figure 2. Alignment of snake venom serine proteases 1-5) Gyroxin-like B2.1, B1.3, B1.4, B1.5, B1.7 from *Crotalus durissus terrificus* (Q58G94, B0FXM1, B0FXM2, EU360953, B0FXM3, respectively). 6) Bilineobin from *Agkistrodon bilineatus* (Q9PSN3). 7) Crotalase from *Crotalus adamanteus*. 8) Ancrod from *Agkistrodon rhodostoma* (P26324). 9) Gyroxin analog from *Lachesis muta muta* (P33589). 10) Acutobin from *Agkistrodon acutus* (Q9I8X2). 11) Calobin from *Agkistrodon caliginosus* (Q91053). 12) KN-BJ from *Bothrops jararaca* (O13069). 13) SVSP-1 Venom serine proteinase from *Crotalus adamanteus* (Q8UUK2). 14) Catroxase-1 from *Crotalus atrox* (Q8QHK3). 15) ACC-C Protein C activator from *Agkistrodon contortrix contortrix* (P09872). 16) CPI enzyme from *Agkistrodon caliginosus* (O42207). 17) Dav-PA from *Agkistrodon acutus* (Q9I8X1). 18) Catroxase-2 from *Crotalus atrox* (Q8QHK2). 19) PA-BJ from *Bothrops jararaca* (P81824). 20) TSV-PA from *Trimeresurus stejnegeri* (Q91516). 21) LV-PA from *Lachesis muta muta* (P84036). 22) Batroxobin from *Bothrops atrox* (P04971). Indicated accession numbers are from Swissprot. The lines indicate the disulfide bonds and the catalytic triad (His, Asp and Ser) are represented by *. Toxin names were indicated in bold followed by snake species in italic and the Swissprot accession numbers were represented in parenthesis.

the core of PARs initiates an intracellular signal transduction pathway, which stimulates phosphoinositide breakdown and cytosolic calcium mobilization [37].

There are four PARs (PAR-1, PAR-2, PAR-3 and PAR-4), PAR-1, PAR-3 and PAR-4 can be activated by thrombin while PAR-2 is activated by trypsin and trypsin-like proteases, but not by thrombin [38]. PAR-1 is important for activation of human platelets by thrombin, but plays no apparent role in mouse platelet activation [39]. The consensus sequence among all the

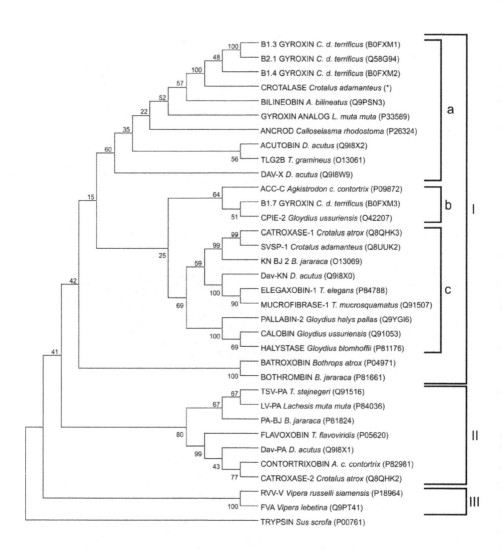

Figure 3. Dendrogram of 34 mature snake venom serine proteases. Toxin names were indicated in bold followed by snake species in italic and the Swissprot accession numbers were represented in parenthesis. (*) Crotalase toxin sequence was based on [19]. All positions containing gaps and missing data were eliminated from the dataset. There were a total of 188 positions in the final dataset. The distance was calculated by number of amino-acid differences. The optimal tree with the sum of branch length = 796.39 and the percentage of replicate trees in which the associated taxa clustered together in the bootstrap test (500 replicates) are shown next to the branches. Fibrinogenolytic (group I), Aα fibrinogenases (subgroup I a), Protein C activator and CPI-enzyme (subgroup I b), kininogenases (subgroup I c), plasminogen activator (group II), factor V activators (group III).

human PAR-1 activating peptides is XFXXR, indicating that the second residue (Phe) and fifth

residue (Arg) are critical for the agonist binding and activation [40].

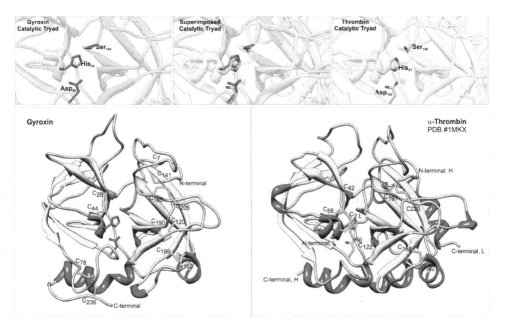

Figure 4. Gyroxin and bovine a-Thrombin (1mkx). The gyroxin homology model was based on crystallographic structures of *Trimeresurus stejnejeri* TSV-PA (1bqy), *Agkistrodon acutus* AaV-SP I (1op0) and *Agkistrodon acutus* AaV-SP II- DAV-PA (1op2). Top: Catalytic tryad superimposition of gyroxin residues (pink) and α-thrombin (blue). Bottom: Two β-barrel domains formed by β-sheets are depicted in yellow and α-helixes in red. Disulfid bridges are depicted in green. The residues from active site are showed in pink to gyroxin molecule and blue to activated thrombin.The C-teminus and N-terminus are indicated to both thrombin heavy (H) and ligth (L) chains and to gyroxin model.

It is known that the inhibitory effects of PAR1 antagonists on platelet aggregation caused by high concentrations of thrombin are limited but can be enhanced by combination with PAR4-blocking antibody, suggesting that simultaneous blockade of PAR1 and PAR4 may provide more effective antithrombotic therapy [41].

An example of snake venom serine protease that acts on coagulation through PAR is gyroxin (serine protease from *C.d.terrificus*) that promotes platelet aggregation through its involvement with PAR 1 and 4 [42]. In fact, a significant inhibition of the maximum platelet aggregation effect induced by gyroxin was observed in the presence of inhibitors of both PAR-1 [SCH79797] and PAR-4 [tcY-NH₂]. PAR-1 inhibitor was effective at concentration of about two orders of magnitude below than that required for PAR-4 inhibitor, and the combination of these two inhibitors were not capable to completely inhibit the platelet aggregation induced by gyroxin [42].

2.2. Molecular biology of SVSPs

Batroxobin (*Bothrops atrox* serine protease, E 3C.4.21.29) is a thrombin-like enzyme derived from *Bothrops atrox, moojeni* venom. In contrast to thrombin which converts fibrino-

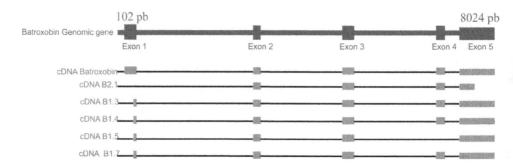

Figure 5. Organization of the batroxobin genomic gene, cDNA of Batroxobin and gyroxin-like B2.1, B.1.3, B1.4, B1.5 and B1.7 from *C.d.terrificus*. Blue boxes denote the location of exons of batroxobin genomic gene (X12747), the blue bars denote noncoding regions of introns. Orange boxes denote the location of exons of cDNA of Batroxobin (J02684) and gyroxin-like B1.3, B1.4, B1.5, B1.7 from *C.d.terrificus* (EU360951; EU360952; EU360953 and EU360954, respectively). In case of clone B2.1, only a partial sequence was obtained (GenBank accession number AY954040). The lack of exon 4 in B1.5 clone is because this clone is truncated by the insertion of a stop codon in translated sequence at the position 472 pb due to the joining of the exon 3 and exon 5.

gen into fibrin by splitting off fibrinopeptides A and B, batroxobin only splits off fibrinopeptide A [43].

Batroxobin gene spans 8 kilobase pairs and contains five exons and its mature form is encoded by exons 2 to 5. The catalytic residues of batroxobin, His-41, Asp-86, and Ser-178, are encoded by separate exons, exons 2, 3, and 5, respectively [44].

The exon/intron organization of the batroxobin gene is different from that of the prothrombin gene but very similar to those of the trypsin and kallikrein genes. These results indicate that batroxobin is not a member of the prothrombin family but one of the trypsin kallikrein family. The snake venom gland is assumed to originate from the submaxillary gland. Therefore, batroxobin is expected to be a member of the glandular kallikrein family [44].

cDNA libraries of snake venom glands have been constructed from various species and several clones encoding SVSPs have been isolated and sequenced. SVSPs are one chain proteins encoded by cDNAs containing an open reading frame (ORF) around 800 bp. The 5'UTRs (5 'untranslated region) are usually short while the 3'UTRs (3'untranslated region) vary in length and may contain more than 1200 nucleotides [3].

Snake venom serine protease are synthesized as zymogens of ~256–257 amino acids with a putative signal peptide of 18 amino acids and a proposed activation peptide of six amino acid residues [3]. In the process of protein export, a central role is played by the signal sequence: an N-terminal segment that somehow initiates export whereupon it is cleaved from the zymogen. Three structurally dissimilar regions have been recognized so far: a positively charged N-terminal region, a central hydrophobic region and a more polar C-terminal region that seems to define the cleavage site [45].

The organization of batroxobin gene, batroxobin cDNA and gyroxin-like B2.1, B1.3, B1.5 and B1.7 [17, 36, 44] are shown in Figure 5.

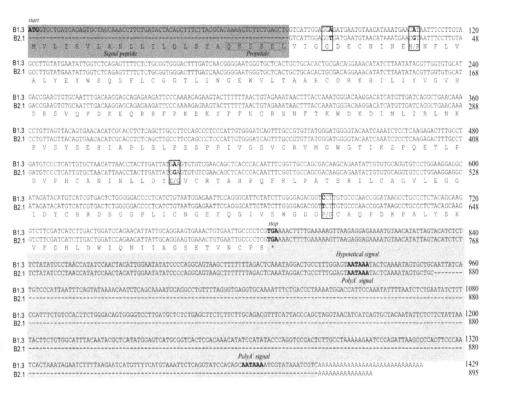

Figure 6. Alignment of nucleotide sequences from gyroxin-like B1.3 (EU360951) and B2.1 (AY954040). Coding region for signal peptide and propeptide is indicated in dark grey. Start codon and stop codons are in bold. The mature coding region is indicated in white. The mutations between B1.3 and B2.1 sequences are indicated by a box line and differences in nucleotides are in bold. Light grey encompasses the 3′ UTR. B1.3 hypothetical poly A+ signal (929–934 bp), B1.3 poly A⁺ signal (1380–1385 bp) and B2.1 poly A⁺ signal (857–862 bp) are in bold. Dashes represent gaps introduced for optimal sequence alignment.

Gyroxin-like B2.1 has a shorter 3′UTR compared with other clones. The lack of exon 4 in gyroxin-like B1.5 is because this clone is truncated by the insertion of a stop codon in translated sequence at the position 472 pb due to the joining of the exon 3 and exon 5 [17].

In Figure 6, the alignment of nucleotide sequences from gyroxin-like B1.3 and B2.1 revealed that clone B1.3 contains two consensus motifs for hypothetical poly(A⁺) signals (5′- AATAAA -3′) at positions 929 and 1380 bp, whereas the B2.1 sequence contains only the first poly(A⁺) signal at the position 857 bp and has a shorter 3′UTR and poly(A⁺) tail [17].

The transcription of mRNA can be related with polyadenilation sites on 3′UTR (3′untranslated region). The presence of short and long 3′UTRs was also described for myogenin, Xmyog U₁ and Xmyog U₂ from *Xenopus laevis* (Xmyog U₂) [46] that contains one and two consensus motifs for a poly(A⁺) signal, respectively. These results suggest the presence of at least, two different poly(A⁺) signals in Xmyog U₂, generating two transcripts with different 3′ ends.

Similarly, the presence of two signals of polyadenylation in gyroxin-like B1.3, suggests that two mRNAs could be transcribed with longer or smaller 3′ UTR. Gyroxin-like B2.1 has only one signal of polyadenilation, showing a shorter 3′UTR than gyroxin-like B1.3 (Figure 5) [17].

2.3. Recombinant serine protease expression

Due to the great biotechnological potential of toxins present in the snakes venoms, many efforts have been made in order to clone and express those toxins in order to study its biological activity. However, the study of their properties is often hampered due to the small amount obtained and the difficulty of getting the animals to extract the poison, and when these are not the case, many toxins require several purification steps that result in a lower final yield. For these reasons, many toxin genes have been isolated, cloned and expressed in heterologous systems. This methodology not only make possible to obtain a large amount of toxins, but also enable amino acids modification by specific mutations in their DNA sequence. Thus, whole molecules may be broken down in order to study the function of its domains [47, 48], as well as amino acid residues may be exchanged for to study its role in substrate binding [49, 50].

2.3.1. Expression of serine protease on prokaryotic cells

Currently, the most used system to express snake toxins has been bacteria. However, the expression of recombinant SVSPs using E. coli as a host may result in expression of insoluble proteins that must be refolded in vitro in order to be activated. Batroxobin, for example, was the first SVSP to be expressed in insoluble form in E. coli and subsequently refolded to yield an active enzyme [51]. The plasminogen activator from T. stejnegeri was expressed in E. coli, but had to undergo a a denaturation-renaturation process in appropriate redox conditions to allow for the correct formation of disulfide bridges. Using an innovative method, a kallikrein-like protease (Tm-5) from the snake venom Taiwan habu (Trimeresurus mucrosquamatus), was expressed in E.coli by placing a polyhistidine-tag linked to an autocatalyzed site based on the cleavage specificity of the serine protease. The autocatalytic cleavage of Tm-5 from the polyhistidine-tagged fusion protein resulted in an active recombinant enzyme [53].

Acutin and mucrosobin, enzymes with fibrinogen-clotting and β-fibrinogenase activities respectively, were successfully expressed in E. coli [54,55]. The expression of a SVSP in E.coli, rCC-PPP, an isoform of cerastocytin from Cerastes cerastes with platelet-aggregating activity was reported [56]. After refolding, the recombinant enzyme showed to be a potent platelet proactivator and to clot fibrinogen.

2.3.2. Expression of serine protease on eukaryotic cells

In general SVSPs are glycosylated and this post-translational modification is important to the toxin activity, besides that, when expressed in E. coli those toxins frequently results in insoluble or inactive forms. Therefore the eukaryotic system such as yeast, mammalian cells and baculovirus expression system in insect cells have been explored, and although the number of works using this systems are small, they are growing substantially, mainly because of its superior refolding machinery and post-translational modifications (e.g. phosphorylation and glycosylation) [57].

The recombinant Haly-PA was successfully expressed using the baculovírus expression system, displayed an indirect fibrinogenolytic activity depending on the presence of plasminogen and cleaved the plasminogen to generate the active plasmin. These results indicate that Haly-PA is a plasminogen activator and displays fibrinogenolytic activity through conversion of plasminogen to plasmin [58].

The recombinant Batroxobin from *Bothrops atrox moojeni* venom –expressed in *Pichia pastoris*, was able to coagulate plasma in a dose dependent manner. However, its molecular weight was higher than the native protein, indicating yeast-type carbohydrate in its structure [59].

The expression of a glycoprotein Gyroxin-like B2.1 from *Crotalus durissus terrificus* venom was reported in COS-7. In order to promote the secretion of this toxin to the culture medium it was fused to the IgK-chain secretion signal peptide at the N terminus [17]. The recombinant Gyroxin expressed in COS-7 cell (Figure 7-Western blot, lane 1) showed the same electrophoretic pattern of the native Gyroxin purified from the venom (Figure 7-Western blot, lane 2). Recombinant Gyroxin-like B2.1 was successfully achieved with esterase activity in the conditioned culture medium, as revealed by immunoblot of secreted protein and standard anti-crotalic serum from Butantan Institute (Figure 7).

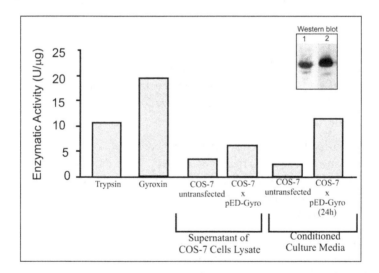

Figure 7. Esterase activity assay of recombinant Gyroxin purified by Benzamidine Sepharose from the supernatant of lysate and conditioned culture medium over 24 h of transfected COS-7 cells with pED-Gyro. Supernatant lysate and conditioned culture medium of untransfected COS-7 cells were used as negative controls. Porcine pancreas Trypsin and purified Gyroxin from *C.d.terrificus* venom were used as positive controls. 1- Western blot of COS-7 cells extract transfected with pED-Gyro., 2- 0.05 mg of Gyroxin purified from *C.d.terrificus* (positive control). The primary antibody was anti-crotalic serum from Butantan Institute and the reaction was detected with secondary antibody conjugated to horseradish peroxidase.

Drug/trade name*	Target and function/treatment	Source
Ancrod (Viprinex)	Fibrinogen inhibitor/stroke	*Agkistrodon rhodostoma* (Malayan pit viper)
Batroxobin (Defibrase)	thrombin and protrhombin inhibitor/acute cerebral infarction, unspecific angina pectoris	*Bothrops moojeni*
Hemocoagulase	thrombin-like effect and thromboplastin activity/ prevention and treatment of haemorrhage	*Bothrops atrox*
Protac/protein C activator	protein C activator/clinical diagnosis of haemostatic disorder	*Agkistrodon contortix contortix* (American copperhead)
Reptilase	diagnosis of blood coagulation disorder	*Bothrops jararaca* (South American lance adder)
RVV-V	Proteolytic activation of factor V	*Daboia ruselli*

Table 1. Clinical applications and diagnostic kits from snake venom serine proteases.

2.4. Therapeutic and diagnostic use

Due to the properties of SVTLEs, they have been extensively investigated over the last decade for potential therapeutic and diagnostic use and some of them are summarized in Table 1, based on [6, 60-62].

In this regard, ancrod [63] from *Calloselasma rhodostoma* venom, whose current brand name is Viprinex™, was approved in 2005 in the fast track program of United States Food and Drug Administration (FDA) for investigating its use in patients suffering from acute ischemic stroke [64].

This program is currently undergoing phase III, where the patients received a one-time, 2-3 hour infusion of ancrod or placebo within six hours of the initial symptom onset of their ischemic stroke, and are then followed for three months to collect information on their functional status. Since then, many research articles about the use of ancrod in ischaemic stroke has been published [63,65-68].

Another thrombin-like enzyme that has been used clinically is Batroxobin (Defibrase®) from *Bothrops atrox* venom. In a randomized clinical trial using this toxin in association with aspirin indicated a reduced rate of restenosis in patients with diabetes undergoing angioplasty for lower-limb ischemia [69]. In another experiment, the combination of batroxobin and tranexamic acid in 80 adolescent patients undergoing scheduled idiopathic scoliosis surgery was able to markedly reduce blood loss and allogeneic blood transfusion [70]. Others trials involving batroxobin include deep vein thrombosis [71] treatment of hyperfibrinogenemia for secondary stroke prevention [72] and acute ischemic stroke [73].

Since SVSPs shortens the bleeding time and clotting time, by promoting coagulation locally at the site of bleeding, combination of enzymes is also employed for the prevention or treatment of hemorrhage such as might be encountered in surgeries. In this regard, hemocoagulase, a mixture of purified enzymes isolated from the venom of *Bothrops atrox* is another example in clinical trials [74,75]. It has two different enzymatic activities, one which promotes blood coagulation by converting prothrombin to thrombin (thromboplastin like enzymes) and the other that causes a direct transformation of fibrinogen to fibrin monomer.

Despite its use in clinical application, some SVSPs have also been explored as a diagnostic tool, mainly because they are not inhibited by heparin and therefore they can be used to test plasma samples containing this anticoagulant or to remove fibrinogen from samples containing heparin. In this context, Reptilase®, a thrombin-like serine protease isolated from the venom of *Bothrops atrox* is used to assess blood fibrinogen and fibrinogen degradation products [76,77]. It is useful to check whether a prolonged thrombin time is caused by the presence of heparin in the sample. However, the reptilase time is rarely performed in isolation and therefore, the results of this test should be considered together with other tests and in particular the thrombin time.

It is important to point out that the name "reptilase" was first described in 1958 [78] for an extract with the fibrinogen clotting activity from the venom of *Bothrops jararaca* and sometimes this term has also been described as a synonymous of "batroxobin" from the venom of *Bothrops moojeni* and *Bothrops atrox*.

Protac®, a serine protease from *Agkistrodon contortrix* venom is another example that has found a broad application in diagnostic practice for the determination of disorders in the protein C (PC) pathway. Unlike thrombin-catalyzed PC activation reaction which requires thrombomodulin as a cofactor, Protac® directly converts the zymogen PC into the catalytically active form which can easily be determined by means of coagulation or chromogenic substrate techniques [79-81].

Due to the capability of a serine protease extracted from Russels´s viper (*Daboia ruselli*) venom (RVV-Vα) to activate Factor V, and since activated factor V is not stable and loses its activity within 20 hours at 37° C, RVV-V has been used to destabilize and selectively inactivate factor V in plasma. Therefore, it has been used to prepare a routine reagent for factor V determination. Studies have demonstrated the ability of the Prothrombinase-induced clotting time (PiCT) assay, which uses RVV-V among its components, to determine activities of both direct and indirect thrombin inhibitors in a linear manner over a wide concentration range [60-62].

While the original native snake venom compounds are usually unsuitable as therapeutics, interventions by medicinal chemists as well as scientists and clinicians in pharmaceutical R&D have made it possible to use snake toxins as therapeutics for multiple disorders based on the available structural and functional information. Therefore, snake venoms, with their cocktail of individual components, have great potential as therapeutic agents for human diseases [6].

2.4.1. Serine protease inhibitors

Most animal species synthesize a variety of protease inhibitors with different specificities, whose function is to prevent unwanted proteolysis. They generally act by unabling access of substrates to the proteases'active site through steric hindrance. Proteases are also involved in various disease states such as the destruction of the extracellular matrix of articular cartilage and bone in arthritic joints is thought to be mediated by excessive proteolitic activity [82]. Among the enzymes involved in extracellular matrix degradation, a few serine proteases (elastase, collagenase, cathepsin G) are able to solubilize fibrous proteins such as elastin and collagen [83,84].

Given the specific recognition by proteases of defined amino acid sequences, it may be possible to inhibit these enzymes when they are involved in pathological processes. Potent inhibitors have the potential to be developed as new therapeutic agents. In vertebrates for example serine protease inhibitors, have been studied for many years and they are known to be involved in phagocytosis, coagulation, complement activation, fibrinolysis, blood pressure regulation. Moreover, some of the protease inhibitors isolated from invertebrate sources are quite specific towards individual mammalian serine proteases. This also offers huge opportunities for medicine. Thus, the development of non-toxic protease inhibitors extracted from invertebrates for *in vivo* application may be quite important [82].

The last decade, drug discovery in leeches has opened the gate for new molecules to treat emphysema, coagulation, inflammation, dermatitis and cancer. Also other invertebrates, such as insects, harvest potential interesting molecules, such as serine protease inhibitors that can be exploited by the medical industry [85].

3. Conclusions

Snake venom serine proteases have several different functions and have found most use in medicine in blood coagulation system. These enzymes are used in several ways as tools in basic research helping to elucidate the relation of structure- function of coagulant proteins and their interactions with platelets or in experimental models of haemostatic alterations.

Some SVSPs have already been found to be a commercial use in coagulation diagnostic and some of them are used either to influence physiological homeostasis or as a form of supportive treatment in haemostatic disorders and micro vascular surgery promoting cicatrization.

Despite the high homology of serine proteases and even sharing the same target, small differences in their amino acids composition may lead significant binding intensity causing differences in their biological effects. Therefore, even isoforms of those molecules in the same organism must be explored. Many animals besides snakes also possess serine proteases that are used for attack or defense purposes, such as scorpions, bees, spiders and even the exotic platypuses which make 26 different kinds of serine proteases [86].

Therefore, the diversity of those toxins is extensive and demand many research to elucidate their function and potential clinical applications

Acknowledgements

Financial support by FAPESP and CNPq.

Author details

Camila Miyagui Yonamine[1], Álvaro Rossan de Brandão Prieto da Silva[2] and
Geraldo Santana Magalhães[3]

1 Department of Pharmacology, Federal University of São Paulo, Brazil

2 Department of Genetic, Butantan Institute, Brazil

3 Department of Immunology, Butantan Institute, Brazil

References

[1] Kornalik F. Toxins affecting blood coagulation and fibrinolysis. In: Shier W. T. and
Meb D. (eds.) Handbook of Toxicology. New York: Marcel Dekker; 1990. p.697-709.

[2] Matsui T, Fujimura Y, Titani K. Snake venom proteases affecting hemostasis and
thrombosis. Biochim. Biophys.Acta 2000;1477(1-2) 146–156.

[3] Serrano S, Mauron RC. Snake venom serine proteinases: sequence homology vs. Substrate specificity, a paradox to be solved. Toxicon 2005;45(8) 1115-1132.

[4] Bell WRJr. Defibrinogenating enzymes. Drugs 1997;54(3) 18–30,discussion 30-1.

[5] Pirkle H. Thrombin-like enzymes from snake venoms: an update inventory. Thromb.
Haemost. 1998;79(3) 675–683.

[6] Koh DCI, Armugan A, Jeyaseelan K. Snake venom components and their applications in biomedicine. Cell. Mol. Life Sci. 2006;63(24) 3030-3041.

[7] Stocker K, Fischer H, Meier J. Thrombin-like snake venom proteinases. Toxicon
1982;20(1) 265-273.

[8] Pirkle H, Theodor I. Thrombin-like venom enzymes: Structure and function. Adv.
Exp. Med.Biol. 1990;281: 165-175.

[9] Serrano SMT, Hagiwara Y, Murayama N, Higuchi S, Mentele R, Sampaio CA, Camargo AC, Fink E. Purification and characterization of a kinin-releasing and fibrinogen-clotting serine proteinase (KN-BJ) from the venom of *Bothrops jararaca* and

molecular cloning and sequence analysis of its cDNA. Eur.J.Biochem. 1998;251(3) 845-853.

[10] Wang YM, Wang SR, Tsai IH. Serine protease isoforms of *Deinagkistrodon acutus* venom: cloning, sequencing and phylogenetic analysis. Biochem.J. 2001;354(Pt 1) 161-168.

[11] Oyama E, Takahashi H. Amino acid sequence of a thrombin like enzyme, elegaxobin, from the venom of *Trimeresurus elegans* (Sakishima-habu). Toxicon 2002;40(7) 959-970.

[12] Mcmullen BA, Fujikawa K, Kisiel W. Primary structure of a protein C activator from *Agkistrodon contortrix contortrix* venom. Biochemistry 1989;28(2) 674-679.

[13] Tokunaga F, Nagasawa K, Tamura S, Miyata T, Iwanaga S, Kisiel W. The factor V-activating enzyme (RVV-V) from Russell's viper venom. Identification of isoproteins RVV-V alpha, -V beta, and -V gamma and their complete amino acid sequences. J. Biol. Chem. 1988;263(33) 17471-17481.

[14] Siigur E, Aaspõllu A, Siigur J. Molecular cloning and sequence analysis of a cDNA for factor V activating enzyme. Biochem. Biophys. Res. Commun. 1999;262(2) 328-332.

[15] Matsui T, Sakurai Y, Fujimura Y, Hayashi I, Oh-Ishi S, Suzuki M, Hamako J, Yamamoto Y, Yamazaki J, Kinoshita M, Titani K. Purification and amino acid sequence of halystase from snake venom of *Agkistrodon halys blomhoffii*, a serine protease that cleaves specifically fibrinogen and kininogen. Eur.J.Biochem. 1998;252(3) 569-575.

[16] Hahn BS, Yang KY, Park EM, Chang IM, Kim YS. Purification and molecular cloning of calobin, a thrombin-like enzyme from *Agkistrodon caliginosus* (Korean viper). J. Biochem. 1996;119(5) 835-843.

[17] Yonamine CM, Prieto-da-Silva ARB, Magalhães GS, Rádis-Baptista G, Morganti L, Ambiel FC, Chura-Chambi RM, Yamane T, Camillo MAP. Cloning of serine protease cDNAs from *Crotalus durissus terrificus* venom gland and expression of a functional Gyroxin homologue in COS-7 cells. Toxicon 2009;54(2) 110-120.

[18] Magalhaes A, Da Fonseca BC, Diniz CR, Richardson M. The complete amino acid sequence of a thrombin-like enzyme/gyroxin analogue from venom of the bushmaster snake (*Lachesis muta muta*).FEBS Lett. 1993;329(1-2) 116-120.

[19] Henschen-Edman AH, Theodor I, Edwards BF, Pirkle H. Crotalase, a fibrinogen-clotting snake venom enzyme: primary structure and evidence for a fibrinogen recognition exosite different from thrombin. Thromb. Haemost. 1999;81(1) 81-86.

[20] Burkhart W, Smith GFH, SU JL, Parikh I, Levine HIII. Amino acid sequence determination of ancrod, the thrombin-like alpha-fibrinogenase from the venom of *Akistrodon rhodostoma*. FEBS Lett. 1992;297(3) 297-301.

[21] Itoh N, Tanaka N, Mihashi S, Yamashina I. Molecular cloning and sequence analysis of cDNA for batroxobin, a thrombin-like snake venom enzyme J. Biol. Chem. 1987;262(7) 3132-3135.

[22] Nikai T, Ohara A, Komori Y, Fox JW, Sugihara, H. Primary structure of a coagulant enzyme, bilineobin, from *Agkistrodon bilineatus* venom. Arch. Biochem. Biophys. 1995;318(1) 89-96.

[23] Zhang Y, Wisner A, Xiong YL, Bon C. A novel plasminogen activator from snake venom. Purification, characterization, and molecular cloning. J. Biol. Chem. 1995;270(17) 10246-10255.

[24] Sanchez EF, Santos CI, Magalhaes A, Diniz CR, Figueiredo S, Gilroy J, Richardson M. Isolation of a proteinase with plasminogen-activating activity from *Lachesis muta muta* (bushmaster) snake venom. Arch. Biochem. Biophys. 2000;378(1) 131-141.

[25] Serrano SMT, Mentele R, Sampaio CAM, Fink E. Purification, characterization, and amino acid sequence of a serine proteinase, PA-BJ, with platelet-aggregating activity from the venom of *Bothrops jararaca*. Biochemistry 1995;34 (21) 7186-7193.

[26] Numeric P, Moravie V, Didier M, Chatot-Henry D, Cirille S, Bucher B, Thomas L. Multiple cerebral infarctions following snakebite by *Bothrops carribbaeus*. Am. J. Trop. Med. Hyg.2002;67(3) 287–288.

[27] Markland FS. Snake venom and the hemostatic system. Toxicon 1998;36(12) 1749-1800.

[28] Nolan C, Hall LS, Barlow GH. Ancrod, the coagulating enzyme from Malayan pit viper (*Agkistrodon rhodostoma*) venom. Methods Enzymol. 1976;45: 205-213.

[29] Stocker K, Barlow GH. The coagulant enzyme from *Bothrops atrox* venom (Batroxobin). Methods Enzymol.1976;45: 214-223.

[30] Markland FS, Damus PS. Purification and properties of a thrombin-like enzyme from the venom of *Crotalus adamanteus* (Eastern diamondback rattlesnake). J. Biol. Chem. 1971;246(21) 6460-6473.

[31] Markland, FS. Crotalase. Methods Enzymol.1976;45: 223-236.

[32] Bjarnason JB, Barish A, Direnzo GS, Campbell R, Fox JW. Kallikrein-like enzymes from *Crotalus atrox* venom. J. Biol. Chem.1983;258(20) 12566-12573.

[33] Stocker K, Fischer H, Meier J, Brogli M, Svendsen L. Characterization of protein C activator Protac from venom of the southern copperhead (*Agkistrodon contortrix*) snake. Toxicon 1987;25(3) 239-252.

[34] Marsh N, Williams V. Practical applications of snake venom toxins in haemostasis. Toxicon 2005;45(8) 1171–1181.

[35] Kini RM.The intriguing world of prothrombin activators from snake venom. Toxicon 2005;45(8) 1133-1145.

[36] Yonamine, CM. Cloning of serine proteases from the venom of rattlesnake Crotalus durissus terrificus and expression of a Gyroxin in mammalian cells. Master of science disertation. IPEN (Nuclear and Energy Research Institute); 2007.

[37] Gratio V, Walker F, Lehy T, Laburthe M, Darmoul D. Aberrant expression of protei-nase-activated receptor 4 promotes colon cancer cell proliferation through a persis-tent signaling that involves Src and ErbB-2 kinase. Int. J. Cancer 2009;124(7) 1517–1525.

[38] Coughlin, SR. How the protease thrombin talks to cells. Proc Natl Acad Sci USA 1999;96(20) 11023–11027.

[39] Kahn ML, Zheng YW, Huang W, Bigornia V, Zeng D, Moff S, Farese RVJr, Tam C, Coughlin SR. A dual thrombin receptor system for platelet activation. Nature 1998;394(6694) 690–694.

[40] Mao Y, Jin J, Kunapuli SP. Characterization of a new peptide agonist of the protease-activated receptor-1. Biochem. Pharmacol. 2008;75(2) 438-447.

[41] Kahn ML, Nakanishi-Matsui M, Shapiro MJ, Ishihara H, Coughlin SR. Protease-acti-vated receptors 1 and 4 mediate activation of human platelets by thrombin. J. Clin. Invest. 1999;103(6) 879–887.

[42] Da Silva JAA, Spencer P, Camillo MAP, de Lima VMF. Gyroxin and its biological ac-tivity: effects on CNS basement membranes and endothelium and protease-activated receptors. Curr. Med. Chem.2012;19(2) 281-291.

[43] Stocker K. *Defibrinogenation* with *thrombin-like snake venom enzymes*. In Markwardt F. (ed) Handbook of Experimental Pharmacology. Berlin: Springer-Verlag; 1978. p. 451-484.

[44] Itoh N, Tanaka N, Funakoshi I, Kawasaki T, Mihashi S, Yamashina I. The complete nucleotide sequence of the gene for batroxobin, a thrombin-like snake venom en-zyme. Nucleic Acids Res.1988;16(21) 10377-10378.

[45] Von Heijne G. Signal sequences.The limits of variation. J. Mol. Biol.1985;184(1) 99-105.

[46] Charbonnier F, Gaspera BD, Armand AS, Van der Laarse WJ, Launay T, Becker C, Gallien CL, Chanoine C. Two myogeninrelated genes are differentially expressed in *Xenopus laevis* myogenesis and differ in their ability to transactivate muscle structural genes. J. Biol. Chem.2002;277(2) 1139–1147.

[47] Assakura MT, Silva CA, Mentele R, Camargo AC, Serrano SM. Molecular cloning and expression of structural domains of bothropasin, a P-III metalloproteinase from the venom of Bothrops jararaca. Toxicon 2003;41(2) 217-227.

[48] Moura-da-Silva AM, Línica A, Della-Casa MS, Kamiguti AS, Ho PL, Crampton JM, Theakston RD. Jararhagin ECD-containing disintegrin domain: expression in Escherichia coli and inhibition of the platelet-collagen interaction. Arch Biochem Biophys. 1999;369(2) 295-301.

[49] Tian J, Paquette-Straub C, Sage EH, Funk SE, Patel V, Galileo D, McLane MA. Inhibition of melanoma cell motility by the snake venom disintegrin eristostatin. Toxicon. 2007;49(7) 899-908.

[50] Sanz L, Chen RQ, Pérez A, Hilario R, Juárez P, Marcinkiewicz C, Monleón D, Celda B, Xiong YL, Pérez-Payá E, Calvete JJ. cDNA cloning and functional expression of jerdostatin, a novel RTS-disintegrin from *Trimeresurus jerdonii* and a specific antagonist of the alpha1beta1 integrin. J Biol Chem. 2005;280(49) 40714-40722.

[51] Maeda M, Satoh S, Suzuki S, Niwa M, Itoh N, Yamashina I. Expression of cDNA for batroxobin, a thrombin-like snake venom enzyme. J. Biochem. 1991;109(4) 632-637.

[52] Zhang Y, Wisner, A, Maroun, R.C, Choumet V, Xiong Y, Bon C. *Trimeresurus stejnegeri* Snake Venom Plasminogen Activator SITE-DIRECTED MUTAGENESIS AND MOLECULAR MODELING. J. Biol. Chem 1997; 272(33) 20531-20537.

[53] Hung CC, Chiou SH. Expression of a kallikrein-like protease from the snake venom: engineering of autocatalytic site in the fusion protein to facilitate protein refolding. Biochem. Biophys. Res. Commun.2000;275(3) 924–930.

[54] Pan H, Du X, Yang G, Zhou Y, Wu X. cDNA cloning and expression of acutin. Biochem. Biophys. Res. Commun.1999;255(2) 412–415.

[55] Guo YW, Chang TY, Lin KT, Liu HW, Shih KC, Cheng SH. Cloning and functional expression of the mucrosobin protein, a beta-fibrinogenase of *Trimeresurus mucrosquamatus* (Taiwan Habu). Protein Exp. Purif.2001;23(3) 483–490.

[56] Dekhil H, Wisner A, Marrakchi N, El Ayeb M, Bon C, Karoui H. Molecular cloning and expression of a functional snake venom serine proteinase, with platelet aggregating activity, from the *Cerastes cerastes* viper. Biochemistry 2003;42(36) 10609–10618.

[57] Butler M. Animal cell cultures: recent achievements and perspectives in the production of biopharmaceuticals. Appl. Microbiol. Bioftechnol. 2005,68(3) 283-291.

[58] Park D, Kim H, Chung K, Kim DS, Yun Y. Expression and characterization of a novel plasminogen activator from *Agkistrodon Halys* venom. Toxicon 1998;36(12) 1807-1819.

[59] You WK, Choi WS, Koh YS, Shin HC, Jang Y, Chung KH. Functional characterization of recombinant batroxobin, a snake venom thrombin-like enzyme, expressed from *Pichia pastoris*. FEBS letters 2004;571(1-3) 63-73.

[60] Korte W, Jovic R, Hollenstein M, Degiacomi P, Gautschi M, Ferrández A. The uncalibrated prothrombinase-induced clotting time test. Equally convenient but more pre-

cise than the aPTT for monitoring of unfractionated heparin. Hamostaseologie. 2010;30(4) 212-6.

[61] Calatzis A, Peetz D, Haas S, Spannagl M, Rudin K, Wilmer M. Prothrombinase-induced clotting time assay for determination of the anticoagulant effects of unfractionated and low-molecular-weight heparins, fondaparinux, and thrombin inhibitors. Am. J. Clin. Pathol. 2008;130(3) 446-54.

[62] Fenyvesi T, Jorg I, Harenberg J. Effect of phenprocoumon on monitoring of lepirudin, argatroban, melagatran and unfractionated heparin with the PiCT method. Pathophysiol. Haemost. Thromb. 2002;32(4) 174-179.

[63] Nolan C, Hall LS, Barlow GH. Ancrod, the coagulating enzyme from Malayan pit viper (Agkistrodon rhodostoma) venom. Methods Enzymol. 1976;45: 205–213.

[64] The Internet Strocke Center. http://www.strokecenter.org/trials/clinicalstudies/asp-ii-ancrod-stroke-program-ancrod-viprinex%E2%84%A2-for-the-treatment-of-acute-ischemic-stroke (accessed 2 July 2012).

[65] Liu S, Marder VJ, Levy DE, Wang SJ, Yang F, Paganini-Hill A, Fisher MJ. Ancrod and fibrin formation: perspectives on mechanisms of action. Stroke 2011;42(11) 3277-3280.

[66] Levy DE, del Zoppo GJ, Demaerschalk BM, Demchuk AM, Diener HC, Howard G, Kaste M, Pancioli AM, Ringelstein EB, Spatareanu C, Wasiewski WW. Ancrod in acute ischemic stroke: results of 500 subjects beginning treatment within 6 hours of stroke onset in the ancrod stroke program. Stroke. 2009;40(12) 3796-3803.

[67] Hyperfibrinogenemia and functional outcome from acute ischemic stroke. del Zoppo GJ, Levy DE, Wasiewski WW, Pancioli AM, Demchuk AM, Trammel J, Demaerschalk BM, Kaste M, Albers GW, Ringelstein EB. Stroke. 2009;40(5) 1687-1691.

[68] Hennerici MG, Kay R, Bogousslavsky J, Lenzi GL, Verstraete M, Orgogozo JM, ESTAT investigators. Intravenous ancrod for acute ischaemic stroke in the European Stroke Treatment with Ancrod Trial: a randomised controlled trial. Lancet. 2006;368(9550) 1871-1878.

[69] Wang J, Zhu YQ, Li MH, Zhao JG, Tan HQ, Wang JB, Liu F, Cheng YS.Batroxobin plus aspirin reduces restenosis after angioplasty for arterial occlusive disease in diabetic patients with lower-limb ischemia. J Vasc Interv Radiol. 2011;22(7) 987-994.

[70] Xu C, Wu A, Yue Y. Which is more effective in adolescent idiopathic scoliosis surgery: batroxobin, tranexamic acid or a combination? Arch. Orthop. Trauma Surg. 2012;132(1) 25-31.

[71] Lei Z, Shi Hong L, Li L, Tao YG, Yong LW, Senga H, Renchi Y, Zhong CH. Batroxobin mobilizes circulating endothelial progenitor cells in patients with deep vein thrombosis. Clin Appl Thromb Hemost. 2011;17(1) 75-79.

[72] Xu G, Liu X, Zhu W, Yin Q, Zhang R, Fan X. Feasibility of treating hyperfibrinogenemia with intermittently administered batroxobin in patients with ischemic stroke/

transient ischemic attack for secondary prevention. Blood Coagul Fibrinolysis. 2007;18(2) 193-197.

[73] Gusev EI, Skvortsova VI, Suslina ZA, Avakian GN, Martynov MIu, Temirbaeva SL, Tanashian MA, Kamchtnov PR, Stakhovskaia LV, Efremova NM. Batroxobin in patients with ischemic stroke in the carotid system (the multicenter study). Zh. Nevrol. Psikhiatr. Im. S. S. Korsakova. 2006;106(8) 31-34.

[74] Lodha A, Kamaluddeen M, Akierman A, Amin H. Role of hemocoagulase in pulmonary hemorrhage in preterm infants: a systematic review. Indian J. Pediatr. 2011;78(7) 838-844.

[75] Kim SH, Cho YS, Choi YJ. Intraocular hemocoagulase in human vitrectomy. Jpn. J. Ophthalmol. 1994;38(1) 49–55.

[76] Funk C, Gmür J, Herold R, Straub PW. Reptilase-R--a new reagent in blood coagulation. Br. J. Haematol. 1971;21(1) 43-52

[77] Van Cott EM, Smith EY, Galanakis DK. Elevated fibrinogen in an acute phase reaction prolongs the reptilase time but typically not the thrombin time. Am. J. Clin. Pathol. 2002;118(2) 263-268.

[78] Blomback B, Blomback M, Nilsson IM. Coagulation studies on reptilase, an extract of the venom from Bothrops jararaca. Thromb. Diath. Haemorrh. 1958;1(1) 76-86.

[79] Green L, Safa O, Machin SJ, Mackie IJ, Ryland K, Cohen H, Lawrie AS. Development and application of an automated chromogenic thrombin generation assay that is sensitive to defects in the protein C pathway. Thromb. Res. 2012 Jan 18. [Epub ahead of print].

[80] Tripodi A, Legnani C, Lemma L, Cosmi B, Palareti G, Chantarangkul V, Mannucci PM. Abnormal Protac-induced coagulation inhibition chromogenic assay results are associated with an increased risk of recurrent venous thromboembolism. J. Thromb. Thrombolysis. 2010;30(2) 215-219.

[81] Stocker K, Fischer H, Meier J. Practical application of the protein C activator Protac from Agkistrodon contortrix venom. Folia Haematol. Int. Mag. Klin. Morphol. Blutforsch. 1988;115(3) 260-264.

[82] Royston D. Preventing the inflammatory response to open heart surgery; the role of aprotinin and other protease inhibitors. Int J Cardiol 1996;53: S11-S37.

[83] Sloane BF, Rozhin J, Johnson K, Taylor H, Crissman JD, Honn KV. Cathepsin B: Association with plasma membrane in metastatic tumor. Proc. Natl. Acad. Sci. USA 1986;83(8) 2483-2487.

[84] Berquin IM; Sloane BF. Cathepsin B expression in human tumors. Adv.Exp Med Biol 1996,389: 281-94.

[85] Clynen E., Schoofs L and Salzet M. A. Review of the Most Important Classes of Serine Protease Inhibitors in Insects and Leeches. Med. Chem. Rev. online 2005,2(3)

197-206. http://www.ingentaconnect.com/content/ben/mcro/2005/00000002/00000003/art00003 (acessed 2 July 2012).

[86] Whittington CM, Papenfuss AT, Locke DP, Mardis ER, Wilson RK, Abubucker S, Mitreva M, Wong ES, Hsu AL, Kuchel PW, Belov K, Warren WC. Novel venom gene discovery in the platypus.Genome Biol. 2010;11(9) R95.

Chapter 8

Molecular Pharmacology and Toxinology of Venom from Ants

A.F.C. Torres, Y.P. Quinet, A. Havt,
G. Rádis-Baptista and A.M.C. Martins

Additional information is available at the end of the chapter

1. Introduction

In the last decades, poisonous animals have gained notoriety since their venoms (secreted or injected) contain several of potentially useful bioactive substances (polypeptide toxins), which are mostly codified by a single gene or, in the case of venom organic compounds, by a given enzymatic route presented in a specialized tissue where the biosynthesis occur – the venom gland.

In this context, in the age of genomic sciences, sequencing the entire genome or portion of it, can be thought as the straightforward step to understand a given venom composition. Particularly because, in many cases, the venom is produced in so small quantities, requiring great challenge (natural and bureaucratic) to obtain biological material for its investigation or the necessity of sacrifice the animal to get samples for analysis by conventional biochemical methods. Genome sequencing allows us the identification of mRNAs, as well as prediction of protein structure and function. In addition, the construction of cDNA libraries is useful to clone, catalog and identify genes, and subsequently express the proteins of interest from these libraries. By this approach, we can have adequate amounts of polypeptide toxins for functional analysis and application, by which otherwise would be difficult to isolate.

According to [1], venoms' complexity in terms of peptide and protein contents, together with the number of venomous species indicate that only a small proportion (less than 1%) of the all bioactive molecules has been identified and characterized to date, and little is known about the genomic background of the venomous organisms. Consequently, if we take into account that nature, operated by evolutionary processes, is the most efficient source of new functional molecules and drug candidates, the study of all species of venomous animals, including small insects, such as those belonging to the order Hymenoptera [2] will be crucial and timely for basic and applied research.

2. Ants biology: Subfamily Ponerinae

Ants (Vespoidea: Formicidae) belong to the insect order Hymenoptera, which includes other important families like Apidae (bees) and Vespidae (wasps) [3]. The family Formicidae consists of approximately 13.000 species of ants, most of them exibiting an advanced and sophisticated social life. With colonies ranging from tens to millions of individuals, a high diversity as well as numerical and biomass dominance in almost every habitat throughout the world, ants form an important component of terrestrial biodiversity, especially in the Neotropical Region, where about 30% of all known ant species are found [4,5]. All ant species possess eusocial habits, the most conspicuous one being the reproductive division of labor, with one to many queens specialized in reproduction, while the more and less sterile, and nonreproductive workers, help the queen(s) reproduction, tending the brood and dealing with all other tasks of the colony like food collection, nest repair, nest and/territory defense [6].

With more than 1000 species distributed in 28 genera, like *Dinoponera* and *Paraponera*, the Ponerinae subfamily is a primitive group of ants mainly found in tropical habitats [4]. It is also one the four major ant groups (Myrmicinae, Formicinae, Ponerinae and Dolichoderinae), all characterized by high species diversity and widespread geographic distribution [4]. *Dinoponera* Roger, 1861 [7] is a strictly Neotropical genus with six known species [5] that are considered the largest ants of the world (3-4 cm in length): *D. australis*Emery, 1910; *D. gigantea*(Perty, 1833); *D. longipes* Emery, 1901; *D. lucida*Emery, 1901; *D. mutica*Emery, 1901; and *D. quadriceps* Santschi, 1921 (Figure 1). Like in other ponerine ants, *Dinoponera* colonies have a poor social organization, with small colonies that are queenless [9, 10]. Contrary to most ant species, all workers of the *Dinoponera* colony are potential reproductives with functional spermatheca. However, only one (sometimes more) worker mate and become the dominant worker with reproductive function that is regularly disputed by subdominant workers [9, 10]. Like most Ponerinae, *Dinoponera* are mostly predatory ants: their common prey are medium size to large arthropods (mainly insects) that they subdue with their sting [11, 12]

Like all Aculeata hymenopterans (Chrysidoidea, Apoidea, Vespoidea), *Dinoponera* ants have a sting apparatus that is located in the last portion of the gaster, and is formed by the sting itself (derived from the ovipositor of more basal hymenopteran groups) along with two associated glands: the Dufour's gland and the venom gland [4,13]. In all ants, the venom gland apparatus typically consists of paired venom secreting tubules that converge into a single convoluted gland (an elongated continuation of the secretory tubule into the venom gland reservoir), which in turn empties into a sac-like reservoir that leads into the sting (in ants with sting) [4](Figure 2). In *D. australis*, it was shown that the convoluted gland has, like the free tubules, a secretory function [14]. The free tubules and convoluted gland are responsible for toxin production [14], which seems to be composed mainly of proteins [4,15]. Furthermore, it was also shown that its morphology and ultrastructural organization presents similarity with the convoluted gland of vespine waps (Vespinae), a fact that supports the hypothesis of a phylogenetic origin of ants from wasp-like ancestors [14].

Figure 1. *Dinoponera quadriceps* (Quinet, Y.P. 2011)

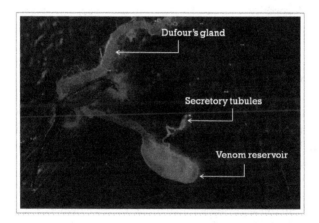

Figure 2. Secretory apparatus from *D. quadriceps* (Quinet, Y.P. 2010)

In solitary Aculeata hymenopterans, and in social bees and wasps, the venom has two main functions: prey capture and defense, respectively [13,16]. In ants, the products from the venom exhibit much higher diversity of biological roles. Particularly In stinging ants, particularly in primitive groups like Ponerinae, the primary function of venom gland products is to serve as injectable offensive or/and defensive agents (to capture prey, fight with competitors or against predators, for example) [13,16]. In more derived functions, the venom gland products are used as defensive (toxic and/or repellent) agents by non-stinging ants that topically apply them on the cuticle of enemies, as in *Crematogaster* or *Monomorium* ants for example. Venom gland products can also serve as chemical communication agents (alarm and recruitment phero-mones, for example) [16,17].

3. Clinical aspects of ants' stings

Many insect stings are associated with local pathophysiological events, characterized by pain, swelling and redness at the sting site for about 1-2 days [18]. The most severe reactions are associated with allergic disorders, presenting neutrophilic and eosinophilic infiltration and specific IgE production [19]. These manifestations are common in accidents with Hymenoptera insects. Most studies that describe the clinical aspects of ant stings reported accidents with ants of the genus *Solenopsis* (Myrmicinae), known as fire ants [20,21,22]. In most serious cases, these accidental encounter with fire ants can promote multiple body rash, seizures, heart failure, and serum sickness nephritis and, more rarely, acute renal failure [23,24].

Accidents with ants of the Ponerinae subfamily are rare or rarely reported. In fact, several concomitant or sequential stings are necessary in order to produce significant clinical symptoms of envenomation, in giant ants, multiple attacks are less probable, since workers have a solitary foraging behavior. However, some of the accidents with giants ants may have medical importance, such as the ones produced by the genus *Paraponera* and *Dinoponera*, popularly known as "true tocandira" and "false tocandira", respectively. Their stings are extremely painful and can cause potentially systemic manifestations such as fever, cold sweats, nausea, vomiting, lymphadenopathy and cardiac arrhythmias [8,25,26]. According to [27,28] the venom of these ants may be neurotoxic for other insects.

4. Venom composition and pharmacological properties

The ant's venoms have been investigated in a relatively small number of species. In the group of stinging ants, the most investigated species belong to the Myrmeciinae, Ponerinae, Pseudomyrmecinae and Myrmicinae subfamilies. They produce aqueous solutions of proteinaceous venoms containing enzymatic and non-enzymatic proteins, free amino-acids and small biologically active compounds like histamine, 5-hydroxytryptamine, acetylcholine, norepinephrine, and dopamine [16,17]. Venoms with proteinaceous components are considered as most primitive and are consequently found in other aculeate hymenopterans like wasps and bees [4,16]. A notable exception to this proteinaceous nature of the venom in ants with sting is found in ants of the genera *Solenopsis* (fire ants) and *Monomorium* (Myrmicinae) that produce alkaloid-rich venoms with few proteins. In the Formicinae ants (ex: *Camponotus*, *Formica*), the sting is no more presented, but the poison gland produces a mixtures of simple organic acids an aqueous solution. Formic acid is presented in concentrations up to 65% along with some peptides and free amino-acids [16,17].

As a member of a group of predatory ants (Ponerinae), it is expected that *Dinoponera* would produces such a kind of proteinaceous venom. However, until now few studies have been done with *Dinoponera* venoms. In two of these studies, which compared venoms of a variety of hymenopterans, the presence of proteins, some with enzyme activities (phospholipase A, hyaluronidase, and lipase), was shown for *D. grandis* (in fact, *D. gigantea*) venom [16,29]. In a more recent study, in which the peptide components from the venom of *D. australis* was investigated, over 75 unique protein components were found with a large diversity of properties ranging in size, hydrophobicity, and overall abundance [30]. The biological effects of several ants' venoms have been attributed to their protein repertoire. As showed by [31] high molecular weight proteins are present in the venom of *Dinoponera australis*. In a comparative evaluation of protein composition of hymenopteran venom reservoirs, proteins with molecular weight ranging from 24 to 75kDa were evidenced [29]. Additionaly, two peptides with less than 10 kDa, as well as proteins with molecular weight ranging from 26-90 kDa were also found in the venom of *Myrmecia pilosula* [32]. The electrophoretic profile of wasps also shows variation in the protein molecular weight, ranging from 5 to 200kDa [33,34], whereas the venoms of bees was shown to range from 2 to108 kDa [35].

5. Pharmacology and therapeutic uses of venom form ants

The first reported case about the therapeutic use of venoms from ants were to treat rheumatoid arthritis. In fact, insects might have components that justify its use in traditional medicine in countries of East Asia, Africa and South America [36]. Lately, several studies of ant venom aimed to demonstrate their beneficial intrinsic properties such as reduction of inflammation, pain relief, improved function of the immune system and liver [37,38].

As the venom from Ponerinae subfamily is composed of a complex mixtures of proteins and neurotoxins [39] we would expected to have several pharmacological properties. Small peptides isolated from *Paraponera clavata* venom, called poneratoxin (PoTx) interfere with sodium channels function and have potential use as a biological insecticide [40,41].

Several distinctive pharmacological activities were demonstrated with peptides isolated from *Pachycondyla goeldii* and *Myrmecia* sp. In one of these works, antimicrobial activity against both Gram positive and Gram negative bacteria was observed [42, 43]. In a recent study [44], it was reported that the venom from *Pachycondila sennaarensis* has a significant antitumor effect on breast cancer cells in a dose and time dependent manner without affecting the viability of non tumor cells. In addition, some studies have also shown the renal effects of Hymenoptera venoms. In fact, in more serious accidents with venoms from wasps and bees acute renal failure generally occurs [45,46, 47, 48].

6. Genomic study of ant venom composition

Since the description of DNA double helix by Francis Crick and James Watson (1953), recombinant DNA technology and genomics revolutionized numerous areas of life science. The comprehension of the biochemical and molecular basis of inheritance had been improved our knowledge about the complexity of all forms of life and the manner how genes and proteins interact to create diversity. The genomic revolution was additionally expanded with the advent of bioinformatic, the 'omic' science (transcriptomic, proteomic, peptidome, metabolomic, glycome) and, presently, system biology.

Collective efforts have been joined to annotate the gene composition of insects. The first complete sequenced genome of insect was from the fruit fly *Drosophila melanogaster*, in 2000, followed by a flurry of activities aimed at sequencing the genomes of several additional insect species. In the field of toxinology, the hymenopterans are receiving special attention due to their behavior and the ability to produce venom.

Up to now, at least 10 ant species had their genomes analyzed and published. The ants whose genomes were sequenced include: the fire ant *Solenopsis invicta* found in South America, United States, China, Taiwan, Australia [49]; the Argentine ant *Linepithema humile* [50], the leaf-cutting ant *Acromyrmex echinator* [51] and *Atta cephalote* [52] found in South America; the red harvester *Pogonomyrmex barbatus* found in North and South America [53], the florida carpenter ant *Camponotus floriandus* from United States; and, the jumper ant *Harpegnatos saltator* from India, Sri Lanka and Southeast Asia [54]. Those ant genomes have provided hundreds of new available nucleotide data.

Apart of a detailed genome analysis, the construction of cDNA libraries from ants' venom glands is an important tool in order to analyze venom composition and discover new molecules that could have biological and pharmacological properties. But an important question arises: why hymenopteran venoms? As we pointed at the beginning of this chapter, there are several reports that hymenopteran venom could have biological properties useful for medical purposes. In this scope, from traditional and modern medicine reports, description can be found not only about clinical manifestation caused by hymenopterans venom, as allergic response, but also the benefits of ant venom to treat disease like rheumatoid arthritis and pain [36].

Genomic and transcriptomic studies of hymenopteran cDNA libraries would provide useful information about their protein constituents. Some of these informations would include signal peptide sequences and the presence of post-translational modifications, which cannot be predicted by the studies of mature proteins. Ants genomic studies have shown a number of substances involved in the biology of these insects, such as: vittelogenins, gustatory and odorant receptors, molecules involved in immune response, as well as metabolic and structural proteins like cytochrome P450.

7. Molecular pharmacology and toxinology of *D. quadriceps* venom

Recently, we have initiated a research project dedicated to investigate the composition, the pharmacological properties, and the transcripts from the venom gland components of *Dinoponera quadriceps*.

Using one-dimensional (SDS-PAGE) electrophoresis (1-DE) to resolve *Dinoponera quadriceps* venom proteins, only eight major large polypeptides (ranging from 15 to 100 kDa) were visualized by Comassie Brilliant Blue (CBB) Staining. The 1-DE and the insensitive method of staining with CBB was not adequate to separate small proteins below 15 kDa and peptides (Figure 3)

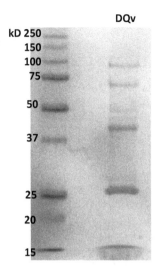

Figure 3. Electrophoretic profile of *Dinoponera quadriceps* total venom (DQv) in one-dimensional SDS-PAGE gel electrophoresis visualized with Comassie Brilliant Blue.

The peptide mass fingerprint (PMF), as well as other proteomic analysis is being conducted and a report will be published elsewhere.

Pharmacological studies have been realized with *Dinoponera quadriceps* venom, particularly, in a system of isolated perfused rat kidney. We now know that at concentrations of approximately 10μg/mL increased urinary flow, glomerular filtration rate and decreased vascular resistance and sodium tubular transport, suggesting a natriuretic and diuretic effect. Furthermore, in studies with renal tubule cells (MDCK - Madin-Darbin Canine Kidney) the same venom induced cell cytotoxicity, on MTT assay (3-(4,5-dimethylthiazol-2-yl)-2,5-diphenyltetrazolium bromide) at a dose and time dependent manner. Interestingly, greater cytotoxicity was observed in the shorter incubation periods, suggesting that the cell culture could recover after a given exposure time. Additional assays have been designed to evaluate the biological and pharmacological activity of purified component of this venom, as well as highlighting the mechanisms related to the observed effects.

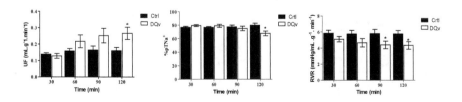

Figure 4. Effect of *D. quadriceps* total venom (DQv) on Urinary flow (UF; A), sodium tubular transport percent(%pTNa; B) and renal vascular resistence (RVR; C). Ctrl=control. Results are expressed as means ± S.E.M., *p<0.05 (ANOVA).

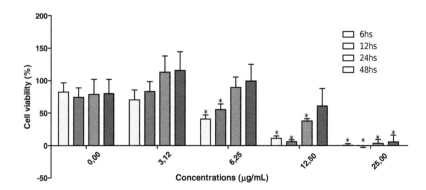

Figure 5. Citotoxicity of *D. quadriceps* total venom on MDCK (Madin-Darbin Canine Kidney) cells culture on MTT assay. Results are expressed as means ± S.E.M., *p<0,05 (ANOVA).

Recently we also demonstrated the neuroprotective activity of *D. quadriceps* venom in models of seizures induced by pentylenetetrazol (PTZ), when administered intraperitoneally. The effect was an increase in latency to first seizure and a tendency to increased latency of death, as well as reduction of lipid peroxidation in the prefrontal cortex of mice [55].

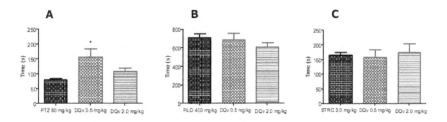

Figure 6. Effects of *D. quadriceps* venom (DQv) on latency of the first seizure in the models of seizure of pentylenetetrazol (PTZ) (A), pilocarpine (PILO) (B) and strychnine (STRC) (C). Results are expressed as means ± S.E.M., *p<0.05 (ANOVA).

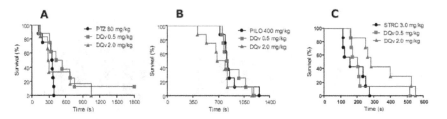

Figure 7. Effects of *D. Quadriceps* total venom (DQv) on latency of death in models of seizure of pentylenetetrazol (PTZ) (A), pilocarpine (PILO) (B) and strychnine (STRC) (C). Results are expressed as means ± S.E.M (n=8), *p<0.05.

A part of proteomic and pharmacological studies, we prepared a *D. quadriceps* venom gland cDNA library to use an EST-strategy to identify the major transcripts expressed in the giant ant venom.we successfully constructed a full-length cDNA library of approximately 20 venom glands from *D. quadriceps*, using In-Fusion SMARTer kit (Clontech, USA). We obtained an efficiency of 1×10^5 cfu/µg of DNA, our medium insert was 700bp and the library was amplified and stored at -80°C. A total of 432 individual ESTs were sequenced by the dideoxy chain termination (Sanger) method. Of these, 125 were undergone to a preliminary analysis through BLASTx. The Tabel 1 and Figure 8(A) shows an overview of the relative abundance of the protein groups.most of the transcripts represent proteins involved in the whole metabolism as transferases, ATP synthase, dehydrogenases, ribosomal proteins, cytocrome c. Those sequences are being annotated for deposit in DNA and protein data bank. A note of caution is that, as in most trancriptome project, a significant number of transcripts showed no similarities with well-known sequences in data bank. These ESTs presents a typical structure of true ORFs (Open Reading Frame), that is start and stop codons, in addition a poly A tail. They were classified as (1) hypothetical proteins with unknown function and (2) cDNA precursors with no hits found. However, by comparing against DNA and protein data the hypothetical proteins showed high similarities with proteins from scorpions (*Opisthacanthus cayaporum*) and others ants, as *Harpegnatos saltator*, *Solenopsis invicta* and *Camponotus floriandus*. The Figure 8(B) represents the percentage of three classification of hits over the total clones analyzed, were probable toxins comprises a significant percentage of ESTs, representing about 34% of messages. Other 37% represents no-significant hits, which give us a number of perspectives to analyze several novel proteins.

Class	Function	% Clones
No hit	Typical ORF with no hits	40.8
DnTx	Mast cell degranulation	28.8
Hypothetical protein	Unknown function	12.0
Antigen like	Allergenic	9.6
Cytocrome c oxidase	Metabolism	1.6
Cytocrome b	Metabolism	1.6
Transferase	Metabolism	2.4
Ionic channel blocker	Toxin	1.6
Ribossomal protein	Structural protein	1.6
Chymotripsin inhibitor	Metabolism	0.8
Dehydrogenase	Metabolism	0.8
ATP synthase	Metabolism	0.8
Phospholipase A1	Enzyme/Toxin	0.8
Bacterial ESTs	Symbionts (?)	4.0
Mitocondrial protein	Metabiolism	0.8

Table 1. Classification of ESTs from *D. quadriceps* venom gland cDNA library on their putative functions.

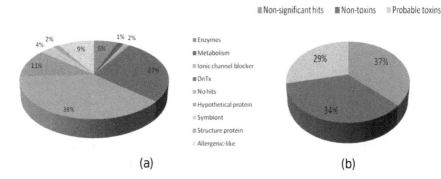

Figure 8. Classification of ESTs from *D. quadriceps* venom gland cDNA library on their putative functions (A). Relative proportion of toxin-encoding, non-toxing encoding and no significant hit ESTs (B).

As a matter of example, the most abundant toxin was dinoponera toxin (DnTx). The dinoponeratoxin whole sequence (accounting for 27% of the total clones analysed) was identified in this cDNA library. Deduced aminoacid sequences (DnTx01 and DnTx02), corresponding to two cDNA isoform precursos, from *D. quadricipes* transcriptome (this work) and three mature venom peptides (DnTx_Da-3105, DnTx_Da-3177 and TX01_DINAS - GenBank accession numbers GI:294863162, GI:294863159 and GI:294863158, respectively) from *D. australis* [30] were aligned with ClustalW software using default parameters (http://www.ebi.ac.uk). DnTx01 and DnTx02 are represented with their respective signal peptides and pro-peptides, in which putative cleavage sites are shown in green and blue, respectively, according to SignalP software (http://www.cbs.dtu.dk/services/SignalP) and proteomic data. In the alignment A is clearly observed that DnTX01 shares high similarity with DnTx_Da-3105 and DnTx_Da-3177, whereas the mature DnTx02 and TX01_DINAS are highly similar to each other (part B).

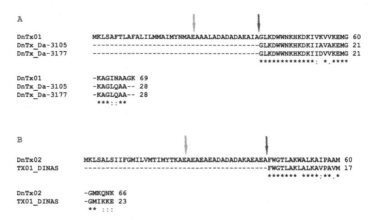

Figure 9. Alignment of dinoponeratoxin precursors and mature peptides from *D. quadricipes* and *D. australis* using ClustalW software (http://www.ebi.ac.uk).

8. Conclusion

Taking into account the information presented in this chapter, a second question arises and should be answered in the near future: "Is there any hymenopteran venom component that could be used as a biotechnological tool?" The majority of works done to discovery new biotechnological tools from hymenopteran venoms were performed using proteomic science analysis, probably because ants apparatus venom is so hard to identify and dissect. Nevertheless, the size of some poneromorph primitive ants may permit subdue these difficulties allowing us to construct a cDNA library and thus opening new perspectives to better understand the biology of ants as well as to analyze the properties of the venom in the search for new molecules with pharmacological and / or biotechnological potential.

Thus, its clear that further work is necessary to understand ant venom, as well venoms from hymenopteran, since several precursors comprises hypothetical and predicted toxins/polypeptides with unknown function. Moreover, a deep functional analysis in the coming period will be made to comprehend the effects presented by total venom and peptides isolated from it.

Acknowledgements

CNPq/CAPES and FUNCAP for financial support.

Author details

A.F.C. Torres[1*], Y.P. Quinet[2], A. Havt[3], G. Rádis-Baptista[4] and A.M.C. Martins[1]

*Address all correspondence to: alba.fabiola@gmail.com

1 Departament of Clinical and Toxicological Analysis, Federal University of Ceara, Fortaleza, Brazil

2 Laboratory of Entomology, State University of Ceara, Fortaleza, Brazil

3 Biomedicine Institute, Department of Physiology and Pharmacology, Federal University of Ceara, Fortaleza, Brazil

4 Marine Science Institute, Federal University of Ceara, Fortaleza, Brazil

References

[1] Ménez A, Stocklin R, Mebs D. 'Venomics' or: The venomous system genome project. Toxicon 2006; 47(3):255-9.

[2]   Harvey AL, Bradley KN, Cochran SA, Rowan EG, Pratt JA, Quillfeldt JA, Jerusalin-sky DA. What can toxins tell us for drug discovery? Toxicon 1998; 36(11):1635-40.

[3]   Gullan PJ, Cranston PS. The insects: an outline of entomology. Wiley-Blackwell:Chi-chester; 2010.

[4]   Holldobler B, Wilson EO. The ants. Belknap Press of Harvard University Press:Cam-bridge;1990.

[5]   Fernández F, Ospina M. Sinopsis de las hormigas de La región Neotropical. In: Fer-nández F (ed.) Introduccíon a lãs Hormigas de La región Neotropical. Instituto de In-vestigación de Recursos Biológicos Alexander Von Humboldt: Colombia; 2003. p49-64

[6]   Wilson EO. The insect societies. Belknap Press of Harvard University Press:Cam-bridge;1971

[7]   Roger J. Die Ponera-artigen Ameisen. (Schluss.)Berliner Entomologische Zeitschrift 1861; 5:1–54.

[8]   Haddad Jr V, Cardoso JLC, Moraes RHP. Description of an injury in a human caused by a false tocandira (*Dinoponera gigantea*, Perty, 1833) with a revision on folkloric, pharmacological and clinical aspects of the giant ants of the genera Paraponera and Dinoponera (sub-family ponerinae). Rev Inst Med Trop 2005; 47(4):235-38.

[9]   Monnin T, Peeters C. Monogyny and regulation of worker mating in the queenless ant *Dinoponera quadriceps*. Animal Behaviour 1998; 55(2):299-306.

[10] Araújo CZD, Jaisson P. Modes de fondation dês colonies chez La fourmi sans reine *Dinoponera quadriceps* Santschi (Hymenoptera,Formicidae, Ponerinae). Actes de Col-loques Insects Sociaux 1994; 9:79-88.

[11] Fourcassié V, Oliveira PS. Foraging ecology of the giant Amazonian ant *Dinoponera gigantea* (Hymenoptera, Formicidae, Ponerinae): activity schedule, diet and spatial foraging patterns. Journal of Natural History 2002; 36:2211-27.

[12] Araujo A, Rodrigues Z. Forraging behavior of the queenless ant *Dinoponera quadriceps* Santschi (Hymenoptera, Formicidae., Neotrop Entomol 2006; 35(2):159-64.

[13] Buschinger A, Maschwitz U..Defensive behavior and defensive mechanisms in ants. In: *Defensive mechanisms in social insects* (H.R. Hermann - Ed.), Praeger, New York; 1984. p95-150

[14] Schoeters E, Billen J. Morphology and ultra structure of the convoluted gland in the ant *Dinoponera australis* (Hymenoptera:Formicidae). Int J Insect Morphol Embryol 1995; 24(3):p323-32.

[15] Siquieroli ACS, Santana FA, Rodrigues RS, Vieira CU, Cardoso R, Goulart LR, Bone-tti AM. Phage display in venom gland in *Dinoponera australis* (HYMENOP-

TERA:FORMICIDAE). J Venom Anim Toxins incl Trop Disc, IX Symposium of the Brazilian Society on Toxinology 2007; 13(1):p291.

[16] Schmidt JO, Blum MS, Overal WL. Comparative enzymology of venoms from stinging Hymenoptera. Toxicon1986; 24(9):p907–21.

[17] Attygalle AB. Morgan ED. Chemicals from the glands of ants. Chemical Society Review 1984;13:p245-78.

[18] Ellis AK, Day JH. Clinical reactivity to insect stings. Current Opinion in Allergy and Clinical Immunol 2005; 5(4):p349-54.

[19] Haddad Jr V. Identificação de enfermidades agudas causadas por animais e plantas em ambientes rurais e litorâneos: auxílio à prática dermatológica. An Bras Dermatol 2009;84(4):p343-8.

[20] Lee EK, Jeong KY, Lyuz DP, Lee YW, Sohn JH, Lim KJ, Hong CS, Park JW. Characterization of the major allergens of Pachycondyla chinensis in ant sting anaphylaxis patients. Clinical and Experimental Allergy 2009; 39:p602-07.

[21] Tankersley MS. The stinging impact of the imported fire ant. Current Opinion in Allergy and Clinical Immunology 2008, 8:p354–59.

[22] Chianura L, Pozzi F. Case Report: A 40-year-old man with ulcerated skin lesions caused by bites of safari ants. Am J Trop Med Hyg 2010; 83(1):p9.

[23] Koya S, Crenshaw D, Agarwal A. Rhabdomyolysis and acute renal failure after fire ant bites. Society of General Internal Medicine 2007;22:p145–47.

[24] Deshazo RD. My journey to the ants. Transactions of the american clinical and climatological association 2009; 120:p85-95.

[25] HADDAD Jr V. Acidentes por formigas. In: Manual de diagnóstico e tratamento de acidentes por animais peçonhentos. Brasília, FNS 2001.p65-66

[26] Cruz Lopez L, Morgan ED. Explanation of bitter taste of venom of ponerine ant *Pachycondyla apicalis*. J Chem Ecol 1997; 23(3):p705-12.

[27] Hermann HR, Blum MS, Wheeler JW, Overal WL, Schmidt JO, Chao J. Comparative anatomy and chemistry of the venom apparatus and mandibular glands in Dinoponera grandis (Guérin) and Paraponera clavata (F.) (Hymenoptera:Formicidae:Ponerinae). Annals of the Entomological Society of America 1984; 77(3):p272-79.

[28] Orivel J, Dejean A. Comparative effect of the venoms of ants of the genus Pachycondyla (Hymenoptera: Ponerinae). Toxicon 2001; 39(2-3):p195-201.

[29] Leluk K, Schmidt J, Jones, D. Comparative studies on the protein composition of hymenopteran venom reservois. Toxicon 1989; 27(1):p105-14.

[30] Johnson SR, Copello JA, Evans MS, Suarez AV. A biochemical characterization of the major peptides from the venom of the giant neotropical hunting ant *Dinoponera australis*. Toxicon 2010; 55(4):p702-10.

[31] Cologna CT, Barbosa DB, Santana FA, Rodovalho CM,Oliveira LA, Brandeburgo MAM. Estudo da peçonha de *Dinoponera australis* – Roger, 1861 (Hymenoptera, Ponerinae). In: V Encontro interno de Iniciação científica, Convênio CNPq/UFU, Uberlândia, 2005.

[32] Wiese MD, Chataway TK, Davies NW, Milne RW, Brown SGA, Gai WP, Heddle RJ. Proteomic analysis of *Myrmecia pilosula* (jack jumper) ant venom. Toxicon 2006; 47(2):p208-17.

[33] Rivers DB, Uckan F, Ergin E. Characterization and biochemical analyses of venom from the ectoparasitic was Nasonia vitripennis (Walker)(Hymenoptera:Pteromalidade). Arch Insect Biochem Physiol 2006; 61(1):p24-41.

[34] Parkinson N, Richards EH, Conyers C, Smith I, Edwards JP. Analysis of venom constituents from the parasitoid wasp Pimpla hypocondriaca and cloning of a cDNA encoding a venom protein.Insect Biochem Mol Biol 2002; 32(7):p729-35.

[35] Zalat S, Nabil Z, Hussein A, Rakha M. Biochemical and haematological studies of some solitary and social bee venoms. Egyptian J of Biology 1999;1:p55-71.

[36] Cherniak EP. Bugs as drugs, Part 1: Insects: the new alternative medicine for the 21st century?. Altern Med Rev 2010; 15(2):p124-35.

[37] Kou J, Ni Y, Li N, Wang J, Liu L, Jiang ZH. Analgesic and anti-inflammatory activities of total extract and individual fractions of Chinese medicinal ants *Polyrhachis lamellidens*. Biol Pharm Bull 2005; 28:p176-180.

[38] Wang CP, Wu YL. Study on mechanism underlying the treatment of rheumatoid arthritis by Keshiling, Zhongguo Zong Yao Za Zhi 2006; 31:155-158.

[39] Sousa PL, Quinet Y, Ponte EL, do Vale JF, Torres AF, Pereira MG, Asseury AM. Venom's antinociceptive property in the primitive ant *Dinoponera quadriceps*. J Ethnopharmacol 2012; 144(1):p213-216.

[40] Lima PRM, Brochetto-Braga MR. Hymenoptera venom review focusing on *Apis mellifera*. J Venomous Animals and Toxins including Tropical Diseases 2003; 9:p149-62.

[41] Duval A, Malécot CO, Pelhate M, Piek T. Poneratoxin, a new toxin from ant venom, reveals an interconversion between two gating modes of the Na channels in frog skeletal muscle fibres. Pflugers Arch 1992; 420(3-4):p239-47.

[42] Szolajska E, Poznanki J, Ferber ML, Michalik J, Gout E, Fender P, Bailly I, Dublet B, Chroboczek J. Poneratoxin, a neurotoxin from ant venom.Structure and expression in insect cells and construction of a bio-insecticide. Europ J Biochem 2004; 271(11):p2127-36.

[43] Orivell J, Redeker V, Caer JP, Krier F, Revol-Junelles A, Longeon A, Chaffotte A, De-jean A, Rossier J. Ponericins, new antibacterial and insecticidal peptides from venom of the ant *Pachycondyla goeldii*. J Biochem Chem 2001; 286(21):p17823-9.

[44] Zelezetsky I, Pag U, Antcheva N, Sahl HG, Tossi A. Identification and optimization of an antimicrobial peptide from the ant venom toxin pilosulin. Biochem Biophys 2005; 434(2):p358-64.

[45] Badr G, Garraud O, Daghestani M, Al-Khalifa MS, Richard Y. Human breast carcino-ma cells are induced to apoptosis by samsum ant venom through an IGF-1-depend-ant pathway, PI3K/AKT and ERK signaling. Cellular Immunology 2012; 273 (1): p10-6.

[46] Daher EF, Oliveira RA, Silva LSV, Silva BEM, Morais TP. Acute renal failure follow-ing bee stings: case reports. Rev Soc Bras Med Trop 2009; 42(2):p209-12.

[47] Grisotto LS, Mendes GE, Castro I, Baptista MA, Alves VA, Yu L, Burdmann EA. Mechanisms of bee venom-induced acute renal failure.Toxicon 2006; 48(1):p44-54.

[48] Loh HH, Tan CH. Acute renal failure and posterior reversible encephalopathy syn-drome following multiple waspstings: a case report. Med J Malaysia 2012; 67(1):p133-5.

[49] Rachaiah NM, Jayappagowda LA, Siddabyrappa HB, Bharath VK.Unusual case of acute renal failure following multiple wasp stings. N Am J Med Sci 2012; 4(2):p104-6.

[50] Wurm Y, Wang J, Riba-Grognuz O, Corona M, Nygaard S, Hunt BG, Ingram KK, Fal-quet L, Nipitwattanaphon M, Gotzek D, Dijkstra MB, Oettler J, Comtesse F, Shih CJ, Wu WJ, Yang CC, Thomas J, Beaudoing E, Pradervand S, Flegel V, Cook ED, Fab-bretti R, Stockinger H, Long L, Farmerie WG, Oakey J, Boomsma JJ, Pamilo P, Yi SV, Heinze J, Goodisman MA, Farinelli L, Harshman K, Hulo N, Cerutti L, Xenarios I, Shoemaker D, Keller L. The genome of the fire ant *Solenopsisinvicta*. Proc Natl Acad Sci U S A 2011; 108(14):p5679-84.

[51] Smith CD, Zimin A, Holt C, Abouheif E, Benton R, Cash E, Croset V, Currie CR, El-haik E, Elsik CG, Fave MJ, Fernandes V, Gadau J, Gibson JD, Graur D, Grubbs KJ, Hagen DE, Helmkampf M, Holley JA, Hu H, Viniegra AS, Johnson BR, Johnson RM, Khila A, Kim JW, Laird J, Mathis KA, Moeller JA, Muñoz-Torres MC, Murphy MC, Nakamura R, Nigam S, Overson RP, Placek JE, Rajakumar R, Reese JT, Robertson HM, Smith CR, Suarez AV, Suen G, Suhr EL, Tao S, Torres CW, van Wilgenburg E, Viljakainen L, Walden KK, Wild AL, Yandell M, Yorke JA, Tsutsui ND. Draft ge-nome of the globally widespread and invasive Argentine ant (*Linepithema humile*). Proc Natl Acad Sci U S A 2011; 108(14):p5673-8.

[52] Nygaard S, Zhang G, Schiøtt M, Li C, Wurm Y, Hu H, Zhou J, Ji L, Qiu F, Rasmussen M, Pan H, Hauser F, Krogh A, Grimmelikhuijzen CJ, Wang J, Boomsma JJ. The ge-nome of the leaf-cutting ant *Acromyrmex echinatior* suggests key adaptations to ad-vanced social life and fungus farming.Genome Res 2011; 21(8):p1339-48.

[53] Suen G, Teiling C, Li L, Holt C, Abouheif E, Bornberg-Bauer E, Bouffard P, Caldera EJ, Cash E, Cavanaugh A, Denas O, Elhaik E, Favé MJ, Gadau J, Gibson JD, Graur D, Grubbs KJ, Hagen DE, Harkins TT, Helmkampf M, Hu H, Johnson BR, Kim J, Marsh SE, Moeller JA, Muñoz-Torres MC, Murphy MC, Naughton MC, Nigam S, Overson R, Rajakumar R, Reese JT, Scott JJ, Smith CR, Tao S, Tsutsui ND, Viljakainen L, Wissler L, Yandell MD, Zimmer F, Taylor J, Slater SC, Clifton SW, Warren WC, Elsik CG, Smith CD, Weinstock GM, Gerardo NM, Currie CR. The genome sequence of the leaf-cutter ant *Atta cephalotes* reveals insights into its obligate symbiotic lifestyle. PLoS Genet 2011; 7(2):e1002007.

[54] Smith CR, Smith CD, Robertson HM, Helmkampf M, Zimin A, Yandell M, Holt C, Hu H, Abouheif E, Benton R, Cash E, Croset V, Currie CR, Elhaik E, Elsik CG, Favé MJ, Fernandes V, Gibson JD, Graur D, Gronenberg W, Grubbs KJ, Hagen DE, Viniegra AS, Johnson BR, Johnson RM, Khila A, Kim JW, Mathis KA, Munoz-Torres MC, Murphy MC, Mustard JA, Nakamura R, Niehuis O, Nigam S, Overson RP, Placek JE, Rajakumar R, Reese JT, Suen G, Tao S, Torres CW, Tsutsui ND, Viljakainen L, Wolschin F, Gadau J.Draft genome of the red harvester ant *Pogonomyrmex barbatus*. Proc Natl Acad Sci U S A 2011; 108(14):p5667-72.

[55] Bonasio R, Zhang G, Ye C, Mutti NS, Fang X, Qin N, Donahue G, Yang P, Li Q, Li C, Zhang P, Huang Z, Berger SL, Reinberg D, Wang J, Liebig J. Genomic comparison of the ants *Camponotus floridanus* and *Harpegnathos saltator*.Science 2010; 329(5995):p1068-71.

[56] Lopes KS, Rios E, Dantas RT, Lima C, Linhares M, Torres AFC, Menezes R, Quinet YP, Havt A, Fonteles MM, Martins AMC. Effect of *Dinoponera quadriceps* venom on chemical-induced seizures models in mice. Rencontres en Toxinologie – Meeting on Toxinology SFET Editions 2011. http://www.sfet.asso.fr (accessed 20 August 2012).

Chapter 9

Discovering the Role of MicroRNAs in Microcystin-Induced Toxicity in Fish

Paweł Brzuzan, Maciej Woźny, Lidia Wolińska and
Michał K. Łuczyński

Additional information is available at the end of the chapter

1. Introduction

MicroRNAs (miRNAs) form a class of endogenously expressed small, non-coding RNAs, that play key roles in the regulation of gene expression of a broad spectrum of biological processes. However, in the field of toxinology, a science of naturally occurring toxins, the relationship between toxicity and microRNA expression is poorly understood. Microcystins (MCs) are potent cyclic peptide hepatotoxins produced by cyanobacteria, which pose a serious threat to aquatic organisms and may also affect human health through the consumption of contaminated waters or food. Although a number of cell physiologic pathways, potential targets for miRNA regulation, are implicated in the response to MCs in animals, no research so far investigated the role for miRNA genes in the mechanism of microcystin (MC)-induced toxicity in fish. The chapter aims to summarize recent achievements of our team in the field, focusing on expression profiling *in vivo* of liver microRNA levels of whitefish (*Coregonus lavaretus*) following MC-LR exposure.

2. Body

2.1. MicroRNAs in fish cells

MicroRNAs (miRNAs) form a class of endogenously expressed small, non-coding RNAs, that play key roles in the regulation of gene expression of a broad spectrum of biological processes. Figure 1 summarizes crucial steps in microRNA processing. MiRNAs are transcribed by RNA polymerases II or III as primary transcripts (pri-miRNAs), which are further processed by the nuclear RNase III enzyme Drosha to stem-loop-structured miRNA precursor molecules (pre-miRNAs). The pre-miRNAs are subsequently transported to the cytoplasm where the RNase III enzyme Dicer cleaves off the double stranded (ds) portion of the hairpin and generates a short-lived dsRNA of about 19–23 nucleotides (nt) in size. The duplex is subsequently unwound and only one strand gives rise to the mature miR-NA, which is incorporated into miRNA-protein complexes (miRNPs) [1-2]. The mature miRNAs binds to partially complementary recognition sequences located in the 3'-untranslated regions (3'-UTRs) of mRNAs and target them for degradation or translational repression (reviewed in [3]).

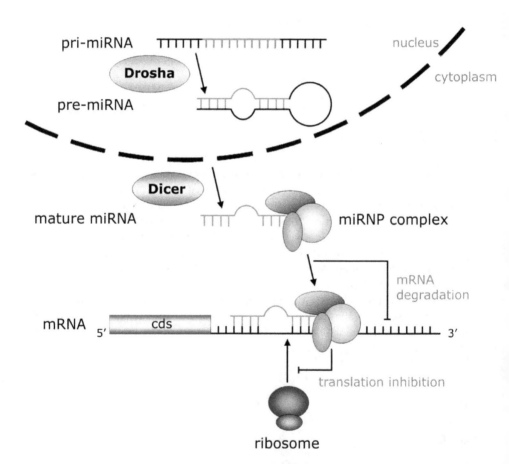

Figure 1. miRNA processing and target recognition. The pri-miRNA is processed by the Drosha enzyme to a stem-loop-structured miRNA precursor molecule (pre-miRNA). The pre-miRNAs is transported to the cytoplasm where the Dicer enzyme cleaves off the double stranded (ds) portion of the hairpin and generates the mature miRNA, which is incorporated into miRNA-protein complexes (miRNPs). The mature miRNA binds to partially complementary recognition sequences on 3'-UTRs of mRNAs and targets them for decay or translational repression.

In metazoans miRNA complementarity to their targets is far from perfect, so one miRNA can bind up to 200 targets, and each mRNA could have recognition sites for more than one miRNA. It is estimated that about 30% of the human protein-coding genes are negatively regulated by miRNA, which suggests that miRNAs are very important regulators of gene expression process [3]. Although specific functions and target mRNAs have been assigned to only a few dozen of miRNAs, much experimental evidence suggests that miRNAs participate in the regulation of a vast spectrum of biological processes. miRNAs control diverse cellular processes including animal development and growth, cell differentiation, signal transduction, cancer, neuronal disease, virus-induced immune defense, programmed cell death, insulin secretion, and metabolism (see [4] and references therein). Understanding of RNA interference (RNAi) has been made possible through a variety of experimental and bioinformatics approaches using different model organisms, including fish [5-6].

To discover aberrantly expressed miRNAs in fish and to determine how altered miRNA function contributes to a disease, new RNAi technologies may be applied (Figure 2). In toxicological studies attention is focused on the relationship between exposure to a chemical and adverse effects it produces in cells, tissues or organisms. So, when a treatment study is carried out small RNA may be collected from a tissue to generate miRNA libraries, from either control or exposed fish. That is the first important step to establish the full repertoire of miRNAs that are differentially regulated in treated fish. Then the miRNA libraries are subjected to massively parallel sequencing, a next generation sequencing technique, which is a combination of emulsion PCR and pyrosequencing [7]. In comparison to microarray analyses, this approach is not limited to previously identified miRNAs and is expected to have superior sensitivity at high sequencing depth. Such approaches have expanded the catalogue of differentially expressed miRNA genes in various fish tissues [6]. The genome-wide screen for regulated miRNAs should yield candidate miRNAs for further profiling (Real Time qPCR) and functional analyses (e.g. Renilla luciferase reporter assay).

As miRNAs regulate many different pathways and orchestrate integrated responses in cells and tissues, it is reasonable to think that they also play key roles in coordinating networks in the poisoned organs. Indeed, there are reports concluding that miRNAs may be key molecules involved in aberrant gene expression in liver cells exposed to hepatotoxic agents, other than MC-LR. For example, Fukushima and co-workers [8] have shown that two well known hepatotoxicants which induce hepatocellular injuries and necrosis, acetaminophen or carbon tetrachloride, were capable of modulating expression of two miRNAs (miR-298 and miR-370) in rats, and that those effects were accompanied by impaired liver metabolism. The observation that miRNAs levels in rat livers were changed by hepatotoxic compounds prompted our team to investigate the role of fish microRNAs in the context of liver-specific MC-LR toxicity.

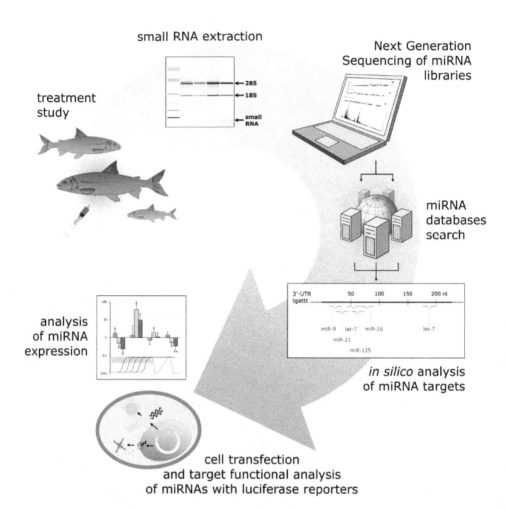

Figure 2. Studying fish miRNAs. MicroRNA discovery has been recently revolutionized by next-generation sequencing. Following ligation of specific linkers to small RNAs (which comprise miRNAs), cDNAs can be produced, which are ideally suited to sequencing using short-read platforms. Databases now offer online catalogues of known microRNAs, which may further be examined for their pathways and functions through a variety of approaches, such as target functional analysis of candidate miRNAs using luciferase reporter assays and miRNA profiling with Real Time PCR.

2.2. Microcystins as potent cyanobacterial toxins

Microcystins (MCs) are potent hepatotoxins produced by cyanobacteria of the genera *Planktothrix*, *Microcystis*, *Aphanizomenon*, *Nostoc*, or *Anabaena*, which have received worldwide concern in recent decades. Mass growths of cyanobacteria, leading to production of blooms, scums and mats, can occur in nutrient-enriched waterbodies (particularly with phosphorus and nitrogen), enhanced by higher temperature and pH values. MCs can be found in lakes, ponds and rivers used for recreational activities as well as in sources for drinking water preparation [9]. In surface waters, concentrations of total MCs (cell-bound and dissolved) measured with ELISA may reach high levels, of up to 1300 µg/l [9], and thus the toxins may pose a threat to aquatic organisms and humans [10]; the World Health Organization recommends 1 µg/l as the maximum acceptable level for microcystin-LR (MC-LR) in drinking water [11]. So far, more than 100 different structural analogues of MCs have been identified, among which MC-LR is one of the most common and abundant [12].

MCs have strong affinity to serine/threonine specific protein phosphatases (PP1 and PP2A), thereby acting as inhibitors of the enzymes [13]. The acute toxicity of MC can be explained by the PP inhibition, which leads to an excessive phosphorylation of cell proteins, to alterations in the cytoskeleton, and a loss of cell shape [14]. Another biochemical feature of MC toxicity is the production of reactive oxygen species (ROS). MC-related ROS generation has been reported using both *in vitro* approaches with different cell lines of fish and mammals [15-16], as well as in a number of *in vivo* studies in rodent liver, heart and reproductive system [17-19]. This process is related to mitochondrial metabolism and it may lead to cell death and to genotoxicity [20]. Oxidative stress caused by MC exposure is believed to be involved in a series of heart, liver and kidney pathologies [19, 21], neurodegenerative effects [22] and embryotoxicity [23].

In recent years, new insights on the key molecules involved in the signal-transduction and toxicity have been reported [24], which highlighted the complexity of the interaction of these toxins with animal cells (Figure 3). Key proteins involved in MC up-take, biotransformation and excretion have been identified, demonstrating the ability of aquatic animals to metabolize and excrete the toxin. After having caused damage to intestinal (or gill) cells these toxins penetrate liver cell membranes through a bile acid carrier. In liver cells MCs inhibit serine/threonine-specific protein phosphatases, PP1 and PP2A, through the binding to them, thus perturbing signaling pathway controlled by the enzymes. The consequences are induction of mitochondrial permeability and loss of mitochondrial membrane potential leading to dysfunction of the mitochondria, induction of reactive oxygen species (ROS), DNA damage (through lowered expression of DNA-PK), and cell apoptosis (through increase Ca2+ levels, CaMKII). MC activity leads to the differential expression/activity of transcriptional factors (e.g. c-myc, p53) and protein kinases (NeK2) involved in the pathways of cellular differentiation, proliferation, tumor promotion activity, and metastasis [25].

2.3. Likely silencing targets in MC-exposed fish cells

In the field of toxinology, a science of naturally occurring toxins, the relationship between toxicity and microRNA expression is poorly understood. However, based on current knowledge about genes involved in the animal cell response on the exposure to environmental stressors, putative targets for miRNA regulation emerge. Genes of transcription factors, *p53* and *mapk* (mitogen activated protein kinases) regulated proto-oncogenes e.g. *c-myc*, that are involved in MC-LR toxicity (Figure 3), are good candidates for tight and robust regulation by microRNAs. The nuclear phosphoprotein p53 is induced in response to cellular stress. It plays a role as a transcriptional trans-activator in DNA repair, apoptosis and tumor suppression pathways. Interestingly, the protein is a substrate of PP2A [26] and therefore its activity is likely to be regulated, in part, by MC-LR. Furthermore, p53 is a regulator of the expression of the anti- and pro-apoptotic genes including members of the BCL-2 family such as *BCL-2* and *BAX*, as well as *CDKN1A*, encoding p21^{Cip1}, which is a cyclin dependent kinase inhibitor (CDKI), an important effector that acts by inhibiting CDK activity in p53-mediated cell cycle arrest in response to various agents. Indeed, we have shown previously that intraperitoneal injection of whitefish, *Coregonus lavaretus*, with MC-LR at subacute dose of 100 µg/kg body weight induced mRNA expression of tumor suppressor p53 and cyclin dependent kinase inhibitor 1 (cdkn1a) in the liver of exposed fish [27]. Interestingly, it was proven in human cell lines that p53 is a transcription factor for some miRNAs, such as miR-34a [7]. miR-34a mediates some of the well-known effects of p53, i.e. cell cycle arrest or apoptosis, and reduced miR-34a levels can serve as a biomarker for any dysfunction along the p53 axis [28]. Yet, its role in controlling miRNA network in fish awaits investigation (Figure 3).

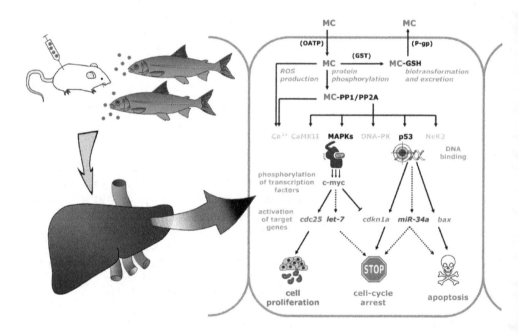

Figure 3. Suggested pathways of MC up-take, toxicity, biotransformation and excretion in vertebrates. Based on the current knowledge, microRNAs (e.g. let-7 or miR-34a) may play roles in MC-LR dependent cell proliferation, cell-cycle arrest or apoptosis.

In the other pathway (Figure 3), mitogen-activated protein kinases (MAPKs) regulate the expression of proto-oncogenes which on the other hand regulate the transcription of genes involved in the growth and differentiation [29]. Expression of MAPKs is mediated by PP2A and are likely to be regulated by MC. The expression of three proto-oncogenes c-fos, c-jun and c-myc were reported to increase in liver, kidney and testis of Wistar rats injected intravenously with MC-LR, with higher levels registered in liver [30]. Expression of these genes suggest that a possible mechanism for the tumor-promoting activity of the toxin could be controlled by MAPKs. Importantly, c-MYC controls expression of let-7 miRNA members by binding to their promoters. The levels of let-7 have been reported to decrease in models of MYC-mediated tumorigenesis, and to increase when MYC is inhibited by chemicals [31]. It is also found that MYC can repress p21^{Cip1} transcription (Figure 1), thereby overriding a p21-mediated cell cycle checkpoint [32].

2.4. miRNA expression in whitefish exposed to MC-LR

In 2008, we began a study of MC-LR induced transcriptional changes in European whitefish, *Coregonus lavaretus* L., a sentinel organism frequently used for pollution monitoring in aquatic systems [27]. To obtain necessary information for the study, full-length cDNA of p53 or cdkn1a of whitefish were determined, using molecular cloning and rapid amplification of cDNA ends (RACE). The *short term* treatment study showed that MC-LR at a dose of 100 μg/kg body weight induced hepatocyte cell DNA fragmentation and up-regulated mRNA expression of p53 and cdkn1a genes in whitefish liver. Interestingly, the elevated transcript levels of both genes were observed only from 48 through the 72 h of exposure, and were accompanied by pathological signs of severe injury of the liver and loss of normal organ functions (elevated levels of blood AspAT AlaAT, and hepatosomatic index; [27]).

Whereas, the above study confirms that MC-LR exposure underlies various acute and chronic effects in fish, it is still little known about aberrant gene expression profiles and molecular pathways involved in the liver of MC-LR challenged organisms. Therefore, to improve our knowledge about adverse effects of MC-LR on hepatocyte cell responses in fish, we performed an initial microRNA study to examine the abundance of 9 selected miRNAs (omy-miR-21, omy-miR-21t, omy-miR-122, omy-miR-125a, omy-miR-125b, omy-miR-125t, omy-miR-199-5a, omy-miR-295, omy-let-7a), in liver samples of whitefish exposed for 24 or 48h to MC-LR at a dose of 100μg/kg body weight [4]. Interestingly, the study showed that MC-LR treatment affected expression levels of two miRNAs, omy-miR-125a (up-regulation) and omy-let-7a (down-regulation) [4].

Following the early demonstration that MC-LR modulates expression of let-7a and miR-125a, in our most recent work [33] we aimed at profiling expression of other 6 miRNAs and 8 mRNAs (Table 1) in the liver of challenged whitefish during the first 48 h after single intraperitoneal injection. From studies on mammals we chose miRNAs which play regulatory roles in pathways of signal transduction (let-7c, [34]; miR-9b, [35]), apoptosis and cell cycle (miR-16a, [36]; miR-21a, [37]; miR-34a, [7]) and fatty-acid metabolism (miR-122, which is a liver specific miRNA, [38]). The selection of mRNA targets (Table 1) was based on their reported aberrant tissue expression on exposure to environmental stressors, and included mRNAs involved in apoptosis and cell cycle (bax, [20]; cas6, cdkn1a, p53, [27]), signal transduction (p-ras, [39]), cellular iron homeostasis (frih, [40]), gene silencing by miRNAs (dcr, [41]), and nucleosome assembly (h2a, [4]). Together with the RNA expression, we analyzed levels of tumor suppressor protein p53 to assess its potential contribution in molecular mechanisms of liver toxicity induced by MCs in fish.

miRNA	Putative biological process*
omy-let-7c	signal transduction
omy-miR-9b	signal transduction
omy-miR-16a	apoptosis, cell cycle
omy-miR-21a	apoptosis, cell cycle
dre-miR-34**	cell cycle, signal transduction
omy-miR-122	fatty-acid metabolism, maintenance of adult liver phenotype
mRNA [gene abbreviation]	Biological process***
bcl2-associated X protein (bax)	apoptosis
caspase 6 (cas6)	apoptosis
cyclin-dependent kinase inhibitor 1a (cdkn1a)	cell cycle
dicer (dcr)	gene silencing by miRNA
ferritin heavy chain (frih)	cellular iron ion homeostasis
histone 2A (h2a)	nucleosome assembly
tumor protein 53 (p53)	apoptosis, cell cycle, signal transduction
HNK Ras –like protein (p-ras)	signal transduction

* based on literature review; see text for details.

** putative *miR-34* gene is present in *Salmo salar* genome; Contig_142190, whole genome shotgun sequence, GenBank ACC. No. AGKD01142167.1, nucleotides from 5978 through 6053.

*** in terms of Gene Ontology Annotation (http://www.ebi.ac.uk/QuickGO).

Table 1. miRNA and mRNA targets selected under study.

Quantifying miRNAs in different tissues is an important initial step in investigating their biological functions. To this end, we determined the expression levels of 6 selected miRNAs in adult whitefish liver using Real-Time qPCR. Prominent expression of miR-122 in the liver of whitefish was observed which is consistent with other data from fish [4,6] and mammals [42]. Variable expression levels of other miRNAs studied in the liver of whitefish corroborated results of previous work on normal human tissues [43], and they are also in agreement with available data on the fish miRNome isolated from rainbow trout [6] and zebrafish [44] miRNA libraries. While the actual expression values of miRNAs can vary by orders of magnitude between whitefish and humans [43], their relative abundance in a particular tissue should tend to be more conserved in evolution. Indeed, the order of individual miRNA abundances in human liver (miR-122 > let-7c ≈ miR-21 ≈ miR16 > miR-34a > miR-9; [43]) held in whitefish as well [33].

Our treatment study [33] identified miRNAs whose expression levels rose (from 2.7-fold for miR-122 to 6.8-fold for let-7c) in MC-LR treated fish, compared to the respective levels in control fish (Figure 4). The increase, which was most apparent at 24 h of the experiment, was correlated with a reduction in the expression of mRNAs: ferritin H (frih) and HNK Ras –like protein (p-ras) and an overexpression of bcl2-associated X protein (bax), cyclin dependent kinase inhibitor 1a (cdkn1a), dicer (dcr), histone 2A (h2a) and p53. Expression of the remaining caspase 6 (cas6) mRNA did not change over 48 h of the treatment. Moreover, exposure to MC-LR did not alter whitefish p53 protein levels [33].

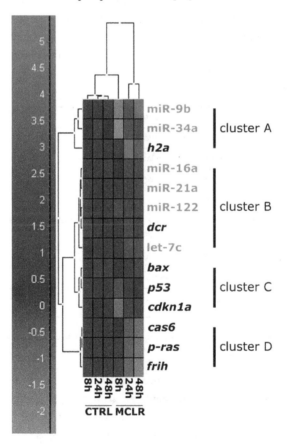

Figure 4. Heat map and hierarchical clustering of differentially expressed genes and miRNAs in MC-LR treated whitefish. Each row represents one gene/miRNA and each column represents a mean of 5 replicates/duration of exposure. Colors represent expression levels of each individual gene/miRNA: red, up-regulation; green, down-regulation. Four distinct clusters (A through D) based on the observed expression profiles could be identified by the analysis. The analysis and visualization were performed using GenEx 5 software (MultiD Analyses AB; Sweden), based on raw expression data from our recent study [33].

The experiment allows one to determine which miRNAs change expression as a group or as a cluster. Genes that function together may define regulatory networks and regulate a common set of regulated genes. Using clustering software, we divided the significantly regulated miRNAs into different groups. In Figure 4 there were four different types of expression profiles among the miRNAs and genes. Some groups showed transient changes in the expression profile (clusters B and C) while others stably increase (cluster A) or decrease (cluster D) during the treatment with MC-LR. Bearing in mind a variety of likely silencing targets for, and the onset of, the aberrant miRNAs expression (Table 2; [33]) it may be concluded that they are involved in diverse molecular pathways, such as liver cell metabolism, cell cycle regulation and apoptosis, and may contribute to the early phase of MC-LR induced hepatotoxicity. Whereas, this argues that at least some of miRNAs listed in Table 2 are good candidates to pursue in future studies, a key to further elucidation of the miRNA role in the toxicity mechanism is the generation of more complete lists of their numbers and expression changes in healthy and challenged fish.

MicroRNA*	Fold change	Reported silencing targets	Reference
let-7c	6.8	Rat sarcoma viral oncogene, RAS	[34]
		Myelocytomatosis viral related oncogene, c-MYC	[45]
miR-9b	4.4	Caudal related homebox protein, CDX2	[35]
miR-34a	4.0	B-cell lymphoma 2, BCL2	[46]
		Myelocytomatosis viral related oncogene, neuroblastoma derived (avian), MYCN	[47]
miR-16a	3.6	B-cell lymphoma 2, BCL2	[36]
miR-122	2.7	Cationic amino acid transporter, CAT-1	[48]

*Only miRNAs which were significantly up-regulated (p<0.05) are included in the column.

Table 2. Reported mammalian silencing targets for differentially expressed miRNAs in MC-LR treated whitefish (100μg/kg body weight) after 24 h of the challenge [33].

On the other hand, the lack of p53 stabilization observed in our study infers the presence of alternate checkpoint mechanisms for deregulated growth signals and/or DNA damage in whitefish cells and may suggest post-transcriptional regulation of *p53*. Indeed, recent work by Liu and coworkers [49] suggest that two checkpoint kinases, ATM and ATR, which act upstream of p53, are promising candidates for the role. Further studies should also reveal if the lack of p53 induction in fish liver following exposure to many compounds known to cause DNA damage and DNA replication defects [49-50], is controlled by the miRNA network, a role it is known to fulfill in other organisms. For example, miR-125b has been previously confirmed to be a negative regulator of p53 in both zebrafish and humans [51].

3. Conclusions

We are only beginning to understand the complexities of miRNA-mediated gene regulatory networks in fish cells. It should be expected that environmental contaminants that have the potential to induce oxidative stress and hypoxia in animal cells, like MCs, will also be agents deregulating miRNA expression. In our initial studies [4, 33] we observed rapid changes in liver microRNA levels of whitefish following MC-LR exposure. Bearing in mind a variety of likely silencing targets for and the onset of the aberrant miRNAs expression observed in the study, one may conclude that they are involved in various molecular pathways and may contribute to the early phase of MC-hepatotoxicity. This argues that studied miRNAs are good candidates to pursue in future studies, however, a key to further elucidation of the miRNA role in the toxicity mechanism will be the generation of more complete lists of their numbers and expression changes in healthy and challenged fish, using next generation sequencing methods (Figure 2). As miRNA field continues to evolve, the new markers should help elucidating a variety of issues intrinsic to MC toxicity. As more profiling studies are performed after MC-LR treatment, and on different model organisms, it might be possible to obtain a miRNA snapshot map, the "core of the MC-LR toxicity connectivity grid". Finally, the revealed miRNA pathways underlying hepatotoxic effects of MC-LR may provide therapeutic targets for a variety of liver diseases.

Acknowledgments

This work was supported by the Polish Ministry of Science and Higher Education (MNiSW), project UWM No. 0809-0801.

Author details

Paweł Brzuzan[1*], Maciej Woźny[1], Lidia Wolińska[1] and Michał K. Łuczyński[2]

1 Department of Environmental Biotechnology, Faculty of Environmental Sciences, University of Warmia and Mazury in Olsztyn, Olsztyn, Poland

2 Department of Chemistry, Faculty of Environmental Management and Agriculture, University of Warmia and Mazury in Olsztyn, Olsztyn, Poland

References

[1] Meister G, Tuschl T. Mechanisms of gene silencing by double-stranded RNA. Nature 2004;431(7006): 343-349.

[2] Ambros V. The functions of animal microRNAs. Nature 2004;431(7006): 350-355.

[3] Zhang B, Wang Q, Pan X. MicroRNAs and their regulatory roles in animals and plants. Journal of Cellular Physiology 2007;210(2): 279-289.

[4] Brzuzan P, Woźny M, Wolińska L, Piasecka A, Łuczyński MK. MicroRNA expression in liver of whitefish (Coregonus lavaretus) exposed to microcystin-LR. Environmental Biotechnology 2010;6(2): 53-60.

[5] Flynt AS, Thatcher EJ, Patton JG. RNA interferences and miRNAs in zebrafish. In: Gaur RK, Rossi JJ. (ed.) Regulation of gene expression by small RNAs. CRC Press; 2009. p149-172.

[6] Salem M, Xiao C, Womack J, Rexroad III CE, Yao J. MicroRNA repertoire for functional genome research in rainbow trout (Oncorhynchus mykiss). Marine Biotechnology 2010;12(4): 410-429.

[7] Tarasov V, Jung P, Verdoodt B, Lodygin D, Epanchintsev A, Menssen A, Meister G, Hermeking H. Differential regulation of microRNAs by p53 revealed by massively parallel sequencing: miR-34a is a p53 target that induces apoptosis and G1-arrest. Cell Cycle 2007;6(13): 1586-1593.

[8] Fukushima T, Hamada Y, Yamada H, Horii I. Changes of micro-RNA expression in rat liver treated by acetaminophen or carbon tetrachloride – regulating role of micro-RNA for RNA expression. Journal of Toxicological Sciences 2007;32(4): 401-409.

[9] Fromme H, Köhler A, Krause R, Führling D. Occurrence of cyanobacterial toxins-microcystins and anatoxin-a–in Berlin water bodies with implications to human health and regulations. Environmental Toxicology 2000;15(2): 120-130.

[10] Carmichael WW. Cyanobacteria secondary metabolites – the cyanotoxins. Journal of Applied Bacteriology 1992;72(6): 445-459.

[11] WHO. Guidelines for Drinking-water Quality, Volume 1. Recommendations, 3rd edition. World Health Organization Publishing 2004; Geneva, Switzerland.

[12] Dietrich D, Hoeger S. Guidance values for microcystins in water and cyanobacterial supplement products (blue-green algal supplements): a reasonable or misguided approach? Toxicology and Applied Pharmacology 2005;203(3): 273-289.

[13] Honkanen RE, Zwiller J, Moore RE, Daily SL, Khatra BS, Dukelow M, Boynton AL. Characterization of microcystin-LR, a potent inhibitor of type 1 and type 2A protein phosphatases. Journal of Biological Chemistry 1990;265(32): 19401-19404.

[14] van Apeldoorn ME, van Egmond HP, Speijers GJA, Bakker GJI. Toxins of cyanobacteria. Molecular Nutrition and Food Research 2007;51(1): 7-60.

[15] Nong Q, Komatsu M, Izumo K, Indo HP, Xu B, Aoyama K, Majima HJ, Horiuchi M, Morimoto K, Takeuchi T. Involvement of reactive oxygen species in Microcystin-LR-induced cytogenotoxicity. Free Radical Research 2007;41(12): 1326-1337.

[16] Pichardo S, Jos A, Zurita JL, Salguero M, Cameán AM, Repetto G. Acute and suba-cute toxic effects produced by microcystin-YR on the fish cell lines RTG-2 and PLHC-1. Toxicology In Vitro 2007;21(8): 1460-1467.

[17] Ding WX, Shen HM, Ong CN. Pivotal role of mitochondrial Ca(2+) in microcystin-induced mitochondrial permeability transition in rat hepatocytes. Biochemical and Biophysical Research Communications 2001;285(5): 1155-1161.

[18] Li Y, Sheng J, Sha J, Han X. The toxic effects of microcystin-LR on the reproductive system of male rats in vivo and in vitro. Reproductive Toxicology 2008;26(3-4): 239-245.

[19] Qiu T, Xie P, Liu Y, Li G, Xiong Q, Hao L, Li H. The profound effects of microcystin on cardiac antioxidant enzymes, mitochondrial function and cardiac toxicity in rat. Toxicology 2009;257(1-2): 86-94.

[20] Žegura B, Filipič M, Šuput D, Lah T, Sedmak B. In vitro genotoxicity of microcystin-RR on primary cultured rat hepatocites and Hep G2 cell line detected by Comet as-say. Radiology and Oncology 2002;36(2): 159-161.

[21] Wei Y, Weng D, Li F, Zou X, Young DO, Ji J, Shen P. Involvement of JNK regulation in oxidative stress-mediated murine liver injury by microcystin-LR. Apoptosis 2008;13(8): 1031-1042.

[22] Feurstein D, Stemmer K, Kleinteich J, Speicher T, Dietrich DR. Microcystin congener- and concentration-dependent induction of murine neuron apoptosis and neurite de-generation. Toxicological Sciences 2011;124(2): 424-431.

[23] Zhao Y, Xiong Q, Xie P. Analysis of microRNA expression in embryonic develop-mental toxicity induced by MC-RR. PLoS ONE 2011;6(7): e22676.

[24] Campos A, Vasconcelos V. Molecular mechanism of microcystin toxicity in animal cells. International Journal of Molecular Sciences 2010;11(1): 268–287.

[25] Zhang XX, Zhang Z, Fu Z, Wang T, Qin W, Xu L, Cheng S, Yang L. Stimulation effect of microcystin-LR on matrix metalloproteinase-2/-9 expression in mouse liver. Toxi-cology Letters 2010;199(3): 377-382.

[26] Li HH, Cai X, Shouse GP, Piluso LG, Liu X. A specific PP2A regulatory subunit, B56γ, mediates DNA damage-induced dephosphorylation of p53 at Thr55. EMBO Journal 2007;26(2): 402-411.

[27] Brzuzan P, Woźny M, Ciesielski S, Łuczyński MK, Góra M, Kuźmiński H, Dobosz S. Microcystin-LR induced apoptosis and mRNA expression of p53 and cdkn1a in liver of whitefish (Coregonus lavaretus L.). Toxicon 2009;54(2): 170-183.

[28] Asslaber D, Piñón JD, Seyfried I, Desch P, Stöcher M, Tinhofer I, Egle A, Merkel O, Greil R. microRNA-34a expression correlates with MDM2 SNP309 polymorphism and treatment-free survival in chronic lymphocytic leukemia. Blood 2010;115(21): 4191-4197.

[29] Gehringer MM. Microcystin-LR and okadaic acid-induced cellular effects: a dualistic response. FEBS Letters 2004;557(1-3): 1-8.

[30] Li H, Xie P, Li G, Hao L, Xiong Q. In vivo study on the effects of microcystin extracts on the expression profiles of proto-oncogenes (c-fos, c-jun and c-myc) in liver, kidney and testis of male Wistar rats injected i.v. with toxins. Toxicon 2009;53(1): 169-175.

[31] Chang TC, Yu D, Lee YS, Wentzel EA, Arking DE, West KM, Dang CV, Thomas-Ti-khonenko A, Mendell JT. Widespread microRNA repression by Myc contributes to tumorigenesis. Nature genetics. 2008;40(1): 43-50.

[32] Gartel AL, Ye X, Goufman E, Shianov P, Hay N, Najmabadi F, Tyner AL. Myc re-presses the p21(WAF1/CIP1) promoter and interacts with Sp1/Sp3. Proceedings of the National Academy of Sciences U.S.A. 2001;98(8): 4510-4515.

[33] Brzuzan P, Woźny M, Wolińska L, Piasecka A. Expression profiling in vivo demon-strates rapid changes in liver microRNA levels of whitefish (Coregonus lavaretus) following microcystin-LR exposure. Aquatic Toxicology 2012;122-123: 188-196.

[34] Johnson SM, Grosshans H, Shingara J, Byrom M, Jarvis R, Cheng A, Labourier E, Reinert KL, Brown D, Slack FJ. RAS is regulated by the let-7 microRNA family. Cell 2005;120(5): 635-647.

[35] Rotkrua P, Akiyama Y, Hashimoto Y, Otsubo T, Yuasa Y. MiR-9 downregulates CDX2 expression in gastric cancer cells. International Journal of Cancer 2011;129(11): 2611-2620.

[36] Cimmino A, Calin GA, Fabbri M, Iorio MV, Ferracin M, Shimizu M, Wojcik SE, Aqei-lan RI, Zupo S, Dono M, Rassenti L, Alder H, Volinia S, Liu CG, Kipps TJ, Negrini M, Croce CM. miR-15 and miR-16 induce apoptosis by targeting BCL2. Proceedings of the National Academy of Sciences U.S.A. 2005;102(39): 13944-13949.

[37] Chan JA, Krichevsky AM, Kosik KS. MicroRNA-21 is an antiapoptotic factor in hu-man glioblastoma cells. Cancer Research 2005;65(14): 6029-6033.

[38] Girard M, Jacquemin E, Munnich A, Lyonnet S, Henrion-Caude A. miR-122, a para-digm for the role of microRNAs in the liver. Journal of Hepatology 2008;48(4): 648-656.

[39] Žegura B, Zajc I, Lah TT, Filipič M. Patterns of microcystin-LR induced alteration of the expression of genes involved in response to DNA damage and apoptosis. Toxi-con 2008;51(4): 615-623.

[40] Woźny M, Brzuzan P, Wolińska L, Góra M, Łuczyński MK. Differential gene expres-sion in rainbow trout (Oncorhynchus mykiss) liver and ovary after exposure to zear-alenone. Comparative Biochemistry and Physiology, Part C: Toxicology and Pharmacology 2012; doi:10.1016/j.cbpc.2012.05.005.

[41] Mishra PK, Tyagi N, Kundu S, Tyagi SC. MicroRNAs are involved in homocysteine-induced cardiac remodeling. Cell Biochemistry and Biophysics 2009;55(3): 153-162.

[42] Ason B, Darnell DK, Wittbrodt B, Berezikov E, Kloosterman WP, Wittbrodt J, Antin PB, Plasterk RH. Differences in vertebrate microRNA expression. Proceedings of the National Academy of Sciences U.S.A. 2006;103(39): 14385-14389.

[43] Liang Y, Ridzon D, Wong L, Chen C. Characterization of microRNA expression profiles in normal human tissues. BMC Genomics 2007;8: 166.

[44] Chen PY, Manninga H, Slanchev K, Chien M, Russo JJ, Ju J, Sheridan R, John B, Marks DS, Gaidatzis D, Sander C, Zavolan M, Tuschl T. The developmental miRNA profiles of zebrafish as determined by small RNA cloning. Genes and Development 2005;19(11): 1288-1293.

[45] Koscianska E, Baev V, Skreka K, Oikonomaki K, Rusinov V, Tabler M, Kalantidis K. Prediction and preliminary validation of oncogene regulation by miRNAs. BMC Molecular Biology 2007;8: 79.

[46] Ji Q, Hao X, Meng Y, Zhang M, DeSano J, Fan D, Xu L. Restoration of tumor suppressor miR-34 inhibits human p53-mutant gastric cancer tumorspheres. BMC Cancer 2008;8: 266.

[47] Wei JS, Song YK, Durinck S, Chen QR, Cheuk AT, Tsang P, Zhang Q, Thiele CJ, Slack A, Shohet J, Khan J. The MYCN oncogene is a direct target of miR-34a. Oncogene 2008;27(39): 5204-5213.

[48] Chang J, Nicolas E, Marks D, Sander C, Lerro A, Buendia MA, Xu C, Mason WS, Moloshok T, Bort R, Zaret KS, Taylor JM. miR-122, a mammalian liver-specific microRNA, is processed from hcr mRNA and may downregulate the high affinity cationic amino acid transporter CAT-1. RNA Biology 2004;1(2): 106-113.

[49] Liu M, Tee C, Zeng F, Sherry JP, Dixon B, Bols NC, Duncker BP. Characterization of p53 expression in rainbow trout. Comparative Biochemistry and Physiology, Part C: Toxicology and Pharmacology 2011;154(4): 326-332.

[50] Rau EM, Billiard SM, Di Giulio RT. Lack of p53 induction in fish cells by model chemotherapeutics. Oncogene 2006;25(14): 2004-2010.

[51] Le MTN, Teh C, Shyh-Chang N, Xie H, Zhou B, Korzh V, Lodish HF, Lim B. MicroRNA-125b is a novel negative regulator of p53. Genes and Development 2009;23(7): 862-876.

Permissions

The contributors of this book come from diverse backgrounds, making this book a truly international effort. This book will bring forth new frontiers with its revolutionizing research information and detailed analysis of the nascent developments around the world.

We would like to thank Dr. Gandhi Radis-Baptista, for lending his expertise to make the book truly unique. He has played a crucial role in the development of this book. Without his invaluable contribution this book wouldn't have been possible. He has made vital efforts to compile up to date information on the varied aspects of this subject to make this book a valuable addition to the collection of many professionals and students.

This book was conceptualized with the vision of imparting up-to-date information and advanced data in this field. To ensure the same, a matchless editorial board was set up. Every individual on the board went through rigorous rounds of assessment to prove their worth. After which they invested a large part of their time researching and compiling the most relevant data for our readers. Conferences and sessions were held from time to time between the editorial board and the contributing authors to present the data in the most comprehensible form. The editorial team has worked tirelessly to provide valuable and valid information to help people across the globe.

Every chapter published in this book has been scrutinized by our experts. Their significance has been extensively debated. The topics covered herein carry significant findings which will fuel the growth of the discipline. They may even be implemented as practical applications or may be referred to as a beginning point for another development. Chapters in this book were first published by InTech; hereby published with permission under the Creative Commons Attribution License or equivalent.

The editorial board has been involved in producing this book since its inception. They have spent rigorous hours researching and exploring the diverse topics which have resulted in the successful publishing of this book. They have passed on their knowledge of decades through this book. To expedite this challenging task, the publisher supported the team at every step. A small team of assistant editors was also appointed to further simplify the editing procedure and attain best results for the readers.

Our editorial team has been hand-picked from every corner of the world. Their multi-ethnicity adds dynamic inputs to the discussions which result in innovative outcomes. These outcomes are then further discussed with the researchers and contributors who give their valuable feedback and opinion regarding the same. The feedback is then collaborated with the researches and they are edited in a comprehensive manner to aid the understanding of the subject.

Apart from the editorial board, the designing team has also invested a significant amount of their time in understanding the subject and creating the most relevant covers. They scrutinized every image to scout for the most suitable representation of the subject and create an appropriate cover for the book.

The publishing team has been involved in this book since its early stages. They were actively engaged in every process, be it collecting the data, connecting with the contributors or procuring relevant information. The team has been an ardent support to the editorial, designing and production team. Their endless efforts to recruit the best for this project, has resulted in the accomplishment of this book. They are a veteran in the field of academics and their pool of knowledge is as vast as their experience in printing. Their expertise and guidance has proved useful at every step. Their uncompromising quality standards have made this book an exceptional effort. Their encouragement from time to time has been an inspiration for everyone.

The publisher and the editorial board hope that this book will prove to be a valuable piece of knowledge for researchers, students, practitioners and scholars across the globe.

List of Contributors

Matheus F. Fernandes-Pedrosa, Juliana Félix-Silva and Yamara A. S. Menezes
Universidade Federal do Rio Grande do Norte, Brazil

Ricardo Bastos Cunha
Bioanalytical Chemistry Laboratory, Division of Analytical Chemistry, Institute of Chemistry, University of Brasília, Brazil

Brasília, Brazil Silvana Giuliatti
Faculty of Medicine of Ribeirão Preto - University of São Paulo, Brazil

Márcia Renata Mortari
Department of Physiological Sciences, Institute of Biological Sciences, University of Brasília, Brazil

Alexandra Olimpio Siqueira Cunha
Department of Physiology, FMRP, University of São Paulo, Brazil

Ana Marisa Chudzinski-Tavassi, Miryam Paola Alvarez-Flores, Linda Christian Carrijo-Carvalho and Maria Esther Ricci-Silva
Laboratory of Biochemistry and Biophysics, Butantan Institute, São Paulo, Brazil

Claudiana Lameu
Departamento de Bioquímica, Instituto de Química, Universidade de São Paulo, Brazil

Márcia Neiva and Mirian A. F. Hayashi
Departamento de Farmacologia, Escola Paulista de Medicina (EPM), Universidade Federal de São Paulo (UNIFESP), Brazil

Camila Miyagui Yonamine
Department of Pharmacology, Federal University of São Paulo, Brazil

Álvaro Rossan de Brandão Prieto da Silva
Department of Genetic, Butantan Institute, Brazil

Geraldo Santana Magalhães
Department of Immunology, Butantan Institute, Brazil

A.F.C. Torres and A.M.C. Martins
Departament of Clinical and Toxicological Analysis, Federal University of Ceara, Fortaleza, Brazil

Y.P. Quinet
Laboratory of Entomology, State University of Ceara, Fortaleza, Brazil

A. Havt
Biomedicine Institute, Department of Physiology and Pharmacology, Federal University of Ceara, Fortaleza, Brazil

G. Rádis-Baptista
Marine Science Institute, Federal University of Ceara, Fortaleza, Brazil

Paweł Brzuzan, Maciej Woźny and Lidia Wolińska
Department of Environmental Biotechnology, Faculty of Environmental Sciences, University of Warmia and Mazury in Olsztyn, Olsztyn, Poland

Michał K. Łuczyński
Department of Chemistry, Faculty of Environmental Management and Agriculture, University of Warmia and Mazury in Olsztyn, Olsztyn, Poland

Printed in the USA
CPSIA information can be obtained
at www.ICGtesting.com
JSHW011428221024
72173JS00004B/713